Confessions
of an Eco-Sinner

Confessions
of an Eco-Sinner

Tracking Down
the Sources of My Stuff

Fred Pearce

Beacon Press, Boston

Beacon Press
25 Beacon Street
Boston, Massachusetts 02108-2892
www.beacon.org

Beacon Press books
are published under the auspices of
the Unitarian Universalist Association of Congregations.

11 10 09 08   8 7 6 5 4 3 2 1

This book is printed on acid-free paper that meets the uncoated paper
ANSI/NISO specifications for permanence as revised in 1992.

Text design by Tag Savage at Wilsted & Taylor Publishing Services

Library of Congress Cataloging-in-Publication Data

Pearce, Fred.
  Confessions of an eco-sinner : tracking down the sources of my stuff /
Fred Pearce.
      p. cm.
  Includes bibliographical references and index.
  ISBN 978-0-8070-8588-2 (acid-free paper) 1. Environmental responsibility.
2. Green movement. 3. Sustainable living. I. Title.

GE195.7P43 2008
333.72—dc22   2008008092

To Sarah

Contents

Part One

Introductions

Footprints 1
Me and My Stuff

We live in a charmed world. If we have money we can buy literally anything. And the majority of us live lifestyles undreamed of only a generation or two ago. One scientist I met recently told me he reckoned that the average household in Europe or North America has so many devices and such a variety of food and clothing that to produce the same lifestyle in Roman times would have required six thousand slaves—cooks, maids, minstrels, ice-house keepers, woodcutters, nubile women with fans, and many more.

I started thinking about that statistic. The scientist's point was that we now rely on machines and cheap energy to do the things that servants would once have done for an élite, while the rest of us went without. But of course it is not that simple. For one thing there are ecological consequences. We gouge out the earth to find the materials to make those machines; and the cheap energy to run them is polluting our planet and warming our climate. And yet many of the servants are still there. Though now, rather than occupying the attics of grand houses, they are spread across the world, growing our food, making our machines, and stitching our clothes.

People talk a lot about carbon footprints. But our personal footprints are much bigger than that. And they are social as well as ecological. The trouble is that in our charmed world we know little about what our footprints are. It all happens so far away. The people and the pollution that sustain us are invisible to us.

I want to change that. My purpose in writing this book was to discover the hidden world that keeps us in the state we have become accustomed to. I have done that by exploring my own personal footprint. I have traveled the world to find out where the cotton in my shirt comes from, the coffee in my mug and the prawns in my curry, the computer on my desk and the phone in my hand, and much else: to discover who grows or mines or makes my stuff, and where that stuff goes after I have finished with it. And to find out whether I should be ashamed of my purchases and their impact on the planet, or whether I should be proud to have contributed to some local economy or given a leg up to some hard-pressed community.

I tried not to pick and choose my journeys too carefully. I simply took those that sounded potentially the most interesting. And certainly I made a point of not changing my way of living to avoid any embarrassments. These are true confessions. And I hope that in tracking down my footprint I will also have tracked down some of yours.

I estimate that I traveled more than 110,000 miles on this journey, visiting more than twenty countries. It took me to the end of my street and the end of my planet, into the African rain forests and the Central Asian deserts, to Bangladeshi sweatshops and Chinese computer factories, to the brothels of Manila and the slums of Rio, to the summits of mountains, the Arctic tundra, the fishing grounds of the Atlantic, and into the bowels of the earth.

Along the way, I met an amazing cross-section of the more than 6 billion people with whom we share this planet. They included a handful of the richest and many of the poorest. It set me wondering as much about the past and future of our species as about the past and future of our planet. And, you may be surprised to learn, it left me with some optimism, about humanity and the huge potential we have to run our world better.

It is easy to be horrified by all the weirdness of humanity. The remote jungle tribes and urban sects, the horrible regimes in North Korea and Myanmar and elsewhere, the fanatics and the fantasists, the warlords and hoodlums, the obscene wealth of the super-rich and the abject state of the super-poor and those struck down by AIDS.

But in my travels I mostly found the commonplace—billions of ordinary people, mostly poor but not starving; mostly ill educated but not uneducated. The resulting book is a kind of chronicle of humanity as represented by the people who grow and make and dispose of the things that I use and consume in my daily life.

So, before I continue with the story, I thought I would stop a moment and describe our strange species, *Homo sapiens,* which has so peremptorily taken over this planet. I drew up a list of things that describe more than a billion people, what we might regard as the mainstream of humanity. I was struck by both what we share and what divides us. So here we are.

A billion of us drive, for instance. Another billion of us have mobile phones. A billion of us can speak English, and another billion eat rice every day. But a billion of us do not have flushing toilets. Some of us share a number of these attributes. But humanity is like a giant version of one of those Venn diagrams, with interlinking circles showing how we differ and how we are the same. A billion of us live in shantytowns, or have access to the Internet, or are Indian, or support a soccer team, or live less than a mile from where we were born, or drink coffee, or have a TV in our home, or cannot support our families. A billion of us have moved from a village to the city, or are Muslims, or have high blood pressure, or use contraceptives regularly, or depend on fish as our main source of protein, or wear sneakers, or are illiterate. A billion of us too are agnostics, or cook with firewood, or have a bicycle, or have heard of Muhammad Ali, or keep chickens, or have a debt with a moneylender, or never consume dairy products, or own T-shirts, or have no electricity.

Collecting these statistics gave me a hugely kaleidoscopic image of *Homo sapiens.* The rainbow race. A billion of us have ridden a bus, or are malnourished, or go to school, or have no running water in our houses. A billion of us are under ten, or are circumcised, or wear jeans, or have a fridge, or smoke cigarettes, or carry the TB bacterium, or eat bananas, or possess some gold, or take annual holidays, or go to the cinema, or pay rent to a landlord, or live on less than a dollar a day. A billion of us drink coke, or will get divorced, or have heard of David Beckham, or are Catholics, or will live beyond our sixtieth year, or are overweight, or are Chinese, or eat bread daily, or get less protein than a Western domestic cat.

But I promised this was going to be a personal journey. So let me introduce myself and, more importantly, my stuff. I am a journalist, in my mid-fifties, with a wife and grown-up children. I live in London and work from home. Most days, I commute to the back bedroom each morning and turn on my computer. But I report about the environment and development around the world. And to do my job, I also travel a lot. This is bad for my carbon footprint, but I really don't believe you can learn about and report on the world by sitting at home and logging on to a virtual reality. The world is much more bizarre and unexpected, and often much more joyful and positive, than you would imagine from reading about it and seeing it on the news. And I found that especially true while following my global footprint. I really did sometimes find that my footprint could be a virtuous one.

In drawing up a list of where I should travel to make this book, I went through my house one day to see where everything came from. It was a shock to find how much of a globalized consumer I have become. When out shopping, I am always the person holding back from a purchase, asking whether I really want it, whether I am being ripped off, or deciding I just can't be bothered to carry it home. And yet, over a lifetime and a couple of trips to the grocery store, it is amazing how much stuff I have, and from how many places.

The food cupboard was an obvious place to start. There were nuts from Brazil and bottles of sparkling water from Scotland; sardines canned in Portugal and anchovies put in a jar in Spain; Italian chopped tomatoes and French mayonnaise; oatmeal from Lancashire and mustard from Dijon; coffee from Tanzania and Marmite from Sussex and soup from a farm in Cambridgeshire. There was mango chutney from India and caster sugar from Mauritius and desiccated coconut from the Philippines.

In some jars at the back, I found dried basil from Egypt and sage from Turkey; arrowroot from Thailand and allspice from Jamaica; lemongrass from Malaysia and galangal from China; mace from Grenada and saffron from Spain. We still haven't got through the bag of white pepper I bought years ago in the market at Kuching in Borneo. Then there were walnuts from China and papadoms from Mumbai, cloves from Indonesia and oregano from Turkey, vanilla from Madagascar and cardamom and ginger from India. Not to mention Spanish olive oil and Italian pasta; Iranian dates and bay leaves from...well, an overgrown bush in our back garden, actually.

In the fruit bowl, I found grapefruit from Florida, mandarins from Argentina, bananas from Costa Rica, nectarines from Spain, and apples from New Zealand. On the counter, I found bread made from Gloucestershire flour, honey made by Mexican bees, and marmalade made by my wife from Seville oranges. There was beer brewed by monks in Belgium, by men in pullovers in Dorset, and by Britain's oldest brewery, amid the hop fields of Kent. There were several bottles of my favorite peaty malt whiskies from the Hebrides, and some untasted spirit from the Netherlands that we have had for two decades. The fridge contained red peppers from Holland and Parmesan from Italy, as well as "the world's largest selling cheese brand," manufactured on the edge of Bodmin Moor from Cornish milk; cucumbers from Essex and a lettuce from Cambridgeshire. The butter and margarine could have come from anywhere.

In the bathroom, I found clove massage oil from Zanzibar, citronella oil from China, and tea tree oil from trees grown in the Australian outback, plus a lot of bottles of various lotions and medicines from the Body Shop that don't declare where they came from.

In my office, I have a rather nice leather briefcase my wife gave me for Christmas one year. It was made in Chennai, India, by Dilip Kapur, a child of the weird religious eco-city known as Auroville. My computer and printer and phone are all made by Taiwanese companies using cheap labor on mainland China; the printer paper came from Slovakia. A random check of the books on my shelves shows they were printed in England and India, Hong Kong and Denmark, the United States and Italy.

Around the house, there are spoils from many of my previous reporting trips. I have a Panama hat from Ecuador and a Russian silk scarf; a Chinese paper dragon, an Alaskan plastic polar bear, and a caiman made from a Brazilian substance rather like rubber; a tiny metal Buddha from India, a job lot of Soviet Lenin badges, a small warthog made of scrap metal from Tanzania, several Japanese fans, and a hand-carved wooden platypus brought back from Queensland by my daughter. In my wardrobe, I found shirts made in Mauritius and Indonesia, Morocco and Cambodia, the United States and Hong Kong. Not to mention a couple of colorful numbers picked up on travels in China and Ghana that I have never worn. Then I found a rather threadbare pullover hand-knitted by a woman I met in the Falkland Islands, a Japanese kimono, Russian army fatigues—and a hat made out of plastic bags by an old woman in a shantytown in South Africa.

In the front room there was an Egyptian mat, an Ashanti stool from Ghana, and a wood spirit carved by a colony of artists in Nigeria. In the back room there was a mat from Jordan, and a mahogany stool we removed from a flat we once rented. A clay recorder and a tambourine are the only survivors of a batch of arty objects I bought for virtually nothing in Mexico in 1984. In the hall I have a cowbell from the Auvergne in France that I use as a dinner gong.

Frankly, our furniture is a bit rudimentary, picked up here and there. But we do have a kitchen table that came from a Victorian workhouse somewhere in Battersea, a gate-leg table, possibly even older, that was a wedding present from my father, and a solid beech desk in my office. We also have a piano. It was made in London in the 1890s by a company called Philips, Cambridge & Co., which is absent from contemporary lists of bona fide piano makers. Our piano tuner said it was not worth tuning anymore, but he tuned it anyway. There seems to be some disagreement about whether the wood is rosewood or mahogany. But there is no doubt that the keys are ivory, probably ripped from central Africa at the height of the greatest slaughter of the elephant ever seen. During the 1890s, Britain imported 550 tons of ivory a year, at a cost of some fifteen thousand animals.

There is other stuff whose legality and morality I wonder about. Somewhere in a cupboard I have two small bags made of sealskin, given to me on a trip to Siberia. We own Burmese teak salad bowls and apartheid-era South African salad servers. Some garden furniture is of dubious provenance, though the stone lion on the lawn is a guaranteed chunk of Cotswold limestone, as carved by my wife's cousin. And the veneer on our kitchen units looks terribly like a piece of illegally logged timber from New Guinea given to me by Greenpeace.

I also got to thinking about the stuff that leaves my house. What happens to the contents of my trash can, for instance? Where do our drains empty? Do the things I put in recycling bins really get recycled?

I can't say I was going to track down the source or destination of all these items. It would be a life's work and I have a life to lead. But I was going in search of some of the most critical to my life—and I hope some of the most interesting. And from the start I was determined to make the journey about people as much as about ecology. Whatever the downsides of globalization, one of the upsides is that it connects us with people from many places. Usually those connections are hidden. But I was off to find them.

Gold

A Lodestone for My Journey

Helmet on. Belt weighed down with emergency oxygen pack. The steel doors shut, a bell sounds from far below, and the cage descends. Very slowly at first. Then suddenly faster. Within fifteen seconds we are traveling at nearly 40 miles an hour down a shaft into the earth. The cage rocks slightly but keeps on descending, longer than you can believe. Somewhere in the dark beside us, another cage shoots past in the opposite direction, hurtling to the surface and full of rock.

Soon we have gone down deeper than the Grand Canyon. It gets warmer. Every 100 yards—every six seconds—the temperature rises by a degree. Down at the bottom of the shaft, almost three miles below the earth's surface and deeper than the ocean floor, the rocks are at more than 120°F. The muggy air is at twice the pressure of the surface air. And radioactive. Scientists have found microbes down here in the dark whose only source of energy is the radiation from the earth's core.

Gradually the lift slows and halts with a reassuring clunk. The doors open and we step out into the earth's crust. I switch on the lamp on my helmet and peer around at the rocks. Three billion years ago, they were the gravelly deposits of a river delta. All sorts of metals were washed down from surrounding mountains that have long since eroded away. And some of those metals accumulated in the gravel beds. One metal in particular accumulated here. And, as a result, this tunnel leads through the heart of what is by a huge margin the richest goldfield on Earth. The West Witwatersrand goldfield in South Africa is also the deepest workplace on the planet. Welcome, says my guide, to Driefontein mine, shaft 7.

I am here to find out where the gold in my wedding ring came from. It is the one thing I never take off; the one thing that came with me every step of the way on my journeys to find my global footprint. As I step out into the Driefontein tunnel, I look again at it. My wife and I bought our bands of gold back in the summer of 1979, in a jeweler's shop on the Strand in central London. We still have the receipts. They cost £50 each. Where did those rings come from? In theory, every piece of gold can be traced back to an individual mine. The gold has its own chemical fingerprint because of the impurities that come with it as it leaves the earth. But in practice, gold in jewelry is usually cast from a range of sources, and my ring defies fingerprinting.

Because most of the 150,000 tons of gold ever mined is still in circulation, the gold on my finger could have been mined in Persia six thousand years ago, or worn by Cleopatra. But ever since gold was discovered in Johannesburg in 1886, and shiploads of Cornish mining engineers came south to exploit them, the mines of West Witwatersrand have produced more than a third of all the gold ever mined, and the rocks here probably still contain more unmined gold than anywhere else.

And on the Witwatersrand, Driefontein is king. The mine that I have entered, some 30 miles southwest of Johannesburg, has been in operation for fifty-five years. During the 1970s, when my ring was cast, Driefontein was the number-one producer in the world's number-one goldfield, yielding up nearly 90 tons of gold a year. In the whole of history, more gold has come out of Driefontein than from anyplace else on Earth. Sometime in the first quarter of 2008, it became the only gold mine ever to have produced more than 3,300 tons of gold. So, as the gates shut behind me, the lift glides on down, and we are left on our own at level 6/41, I can be fairly certain that this hole in the ground is where much of the gold in my ring came from.

We follow some narrow rail tracks away from the shaft and deeper into the mine: tracks that take the miners in each day, and carry out the rock they blast from the gold face. Ducking wooden pit props and electric cables and pipes carrying water and fresh air, I head toward the distant sound of drilling.

The earth's crust here is a maze of different tunnels threading out from each level of the lift shaft. There are fewer than a dozen people in this tunnel today. But at any one time, eight thousand men are working underground in Driefontein alone. On the whole West Witwatersrand goldfield, there are some sixty thousand miners underground. A vast infrastructure is needed just to keep the miners alive. Huge amounts of water naturally course through the mine. Sometimes that water bursts from the dolomite rocks above and into the tunnels. Past floods have threatened to drown the mine forever. To keep the tunnels dry, more than 10 million gallons are pumped out of the mine every day. And as the water is removed, air is sent down to keep the tunnels cool enough to work in. Driefontein has a refrigerator the size of a hangar dedicated to this task.

We continue walking down the tunnel, past a sign warning about a recent death in a neighboring mine. These are dangerous places. There is a constant risk of rockfalls. All along the tunnels wooden pit props, metal jacks, and miles of wire mesh hold up the roof and prevent the walls from falling in. But there are other hazards. Four years ago, here in shaft 7, four miners died after a fire started among the old pit props of some distant abandoned tunnel. Smoke filled the tunnels and more than one hundred miners survived only because they took refuge in air-sealed "refuge chambers." It's like a separate country down here. The miners and supervisors, black and white, have their own language, called Fanakalo. It contains about a thousand words, mostly commands and exhortations and warnings, culled from the numerous African languages spoken by the miners, along with bits of English and Afrikaans. This is the common language of survival. "It's dangerous down there," a geologist told me before we took to the lift. "We really don't know how the rocks behave. People die when things go wrong."

This is an underworld in every sense. Groups of illegal miners smuggle themselves down in the freight lifts. They stay below ground for months at a time, armed with pistols and homemade grenades, chiseling out ore and then grinding it and extracting the gold using mercury, just as poor gold miners do at the surface. Rob Chaplin, the operations manager for the western half of Driefontein, says his supervisors often find, stashed away underground, the balls of mercury used in the crude processing. Syndicates are reckoned to smuggle up to 44 tons of gold out of Witwatersrand every year. This is around a tenth of the total gold taken from the mines. The tunnels are simply impossible to police, Rob says. The surrounding hills are so riddled with old mine shafts and tunnels that you could walk to Johannesburg—a distance of some 30 miles—without once going to the surface.

Since the end of the apartheid era, conditions have improved somewhat for the official miners. At Goldfields, the company that runs Driefontein, the all-male dormitories remain, but there are also houses for conjugal visits. And many miners live in local communities or squatter settlements. Some can make it a modest way up the management chain. There are even a few women working underground, though the white supervisors complain that they strip off to their waists, which is distracting.

Most of the miners have come a long way to take these jobs. They are recruited by agents in their villages in Mozambique or Lesotho or distant parts of rural South Africa—just as they were during the apartheid era. The money they send home is often the mainstay of their communities. Miners' remittances make up some 60 percent of Lesotho's GDP.

Every day, the Driefontein miners leave their company hostels and squatter settlements at 4 a.m. to throng the lifts that hurtle into the earth down one of the mine's eight operational shafts. They fan out along the numerous tunnels at more than twenty different levels. Once at the face, they spend hours with drills the size of gun emplacements, making holes in the hard rock. Then they insert dynamite into the holes. At 5 p.m. every day, the dynamite is exploded all across the mine. The seismic forces sometimes cause high-pressure cave-ins. Then, after the smoke has cleared and any rockfalls have been checked, the night shift goes in to drag out the rubble. More than 20,000 tons of rock come out of Driefontein every day—over 2.5 tons for every miner. In it, invisible to the naked eye, are about 220 pounds of gold—5 grams of gold for every ton of rock brought to the surface.

At the surface, the rock is milled and mixed with cyanide to dissolve the gold, which can then be released from the cyanide by adding one of a range of other chemicals such as zinc. Across the Witwatersrand bush, reservoirs holding the resulting waste ooze cover 155 square miles. Eventually, the rock is transformed into recognizable bars of 85 percent pure gold—before going off to the national Rand Refinery in Johannesburg, where it is refined to 99 percent purity.

So it took more than 2 tons of rock, blasted from the earth's crust, hauled more than a mile to the surface, ground up and treated, to provide enough of the metal for my ring, which weighs less than an ounce. On top of that, making my ring required 5.5 tons of water, more than 30 tons of air pumped underground to keep the mine cool, enough electricity to run a large house for several days—and about ten man-hours of labor. The endeavor behind a simple £50 ring is hard to fathom until you have been down a mine.

The miners, needless to say, see little of the wealth that comes out of these mines. In 2006, while his drillers took home £140 a month, the Goldfields CEO Ian Cockerill was paid almost £50,000 a month. And many of the miners will not live long. The men are far from home and have sex with a lot of different women in the nearby squatter settlements. HIV is widespread in South Africa, and in the mining areas prevalence rates among the men and the female sex workers are as high as 35 percent. Goldfields offers voluntary HIV testing and hands out drugs to its workers, though not to the prostitutes. I met Stella Ntimbane, head of the company's AIDS operation, as she emerged from a testing tent in the compound of one of Driefontein's residential areas. She took gruesome but genuine pride at having uncovered six hundred HIV-positive men in the previous two months, and nine more that morning. "They are getting treatment, and hopefully soon they will be back at the rock face," she says. But ultimately many of them will go back to their villages to die.

For several years from the late 1990s, world gold prices slumped. High-cost mines like Driefontein seemed destined to close. But gold prices soared again in 2005 and now the push is on to extract more and more of the precious metal while prices stay high. Billions of dollars are being bet on that eventuality. Goldfields plans to excavate yet deeper at Driefontein. Shaft 9 is set to go down a farther third of a mile or so, to give the mine another twenty years of life.

It is a miracle these shafts are open at all. Shaft 7, the shaft I have descended, has already been mined once. With me below ground is the mine's chief geologist, Melville Haupt. He has masterminded a return to the abandoned tunnels and galleries to mine more gold by taking out the pillars that were left behind last time to hold up the roof, and to penetrate the areas of difficult geology and fractured rocks previously deemed unsafe. "By rights this place should have shut seven years ago, but Mel just keeps finding more gold," says Chaplin. But Mel is betting his skills as a mine geologist against potential disaster.

Crouching, I follow a narrow passage up to see some old workings. I slip and grab a piece of wood above me. An old pit prop comes away in my hand. Guiltily, I remember an instruction given when we were coming down in the shaft: "Don't grab anything above you. If you slip, just fall." I feel lucky to have survived my indiscretion. They claim to have got through a million man-shifts here since the last fatality. So it would have been a shame to damage the record.

Gold is incorruptible. Since the earliest records, humans have worn gold and hidden gold, bought and sold gold, worshipped gold, fought for gold, and displayed it as a symbol of power. It pervades our language. Some of us are gold diggers, while others have hearts of gold. We have the Midas touch, strike gold, make a pile, and find the golden goose. We pay with gold cards, follow the yellow brick road—or simply watch how things pan out. We look back to a golden age.

Gold is soft and malleable, yet totally indestructible. Nothing in nature attacks it. It is ideal for filling teeth, and for plating critical connections on computer circuit boards. But these are sideshows. Most of the gold ever mined has had no purpose other than its beauty and as a source of value. Gold mainly excites passion. Humans panned for gold before they worked iron or made bronze—perhaps even before they tilled the soil. Gold diggers were certainly among the first people not to have to hunt or grow their own food. Others would do these jobs for them, in return for gold. Gold thus became both the symbol and the substance of wealth.

From the first Egyptian pharaohs and ancient Celtic kings of England, it has been the ultimate form of saving. The first known

European gold jewelry dates from seven thousand years ago. For a thousand years, the ancient Egyptians, for whom gold represented the sun god, maintained a mining colony far away in modern-day southern Sudan to secure their gold. Tutankhamen was found buried in a gold coffin weighing more than 220 pounds. Ancient Britons were mining gold four thousand years ago. The Romans crossed the Sahara for gold. Macedonian gold mines funded Alexander the Great's conquest of the world.

Alchemy was the great science of the Middle Ages; Isaac Newton spent more time trying to turn base metal into gold than he did researching the laws of nature. Later, Europeans colonized the New World for gold, and spent decades searching for El Dorado. The gold rushes of California and Australia and South Africa and the Klondike globalized the world's economy in the nineteenth century. Gold smuggling, more than oil, made the fabulous wealth behind the modern megacity of Dubai.

The gold standard was for a long time the guarantor of the world's currencies; and every national treasury still keeps a store. A third of all the world's gold is locked up in the vaults of various banks and private investment houses. There is Fort Knox, of course. And vaults beneath the Bank of England in the city of London store the bullion reserves of more than seventy countries—hundreds of times more gold than is contained in the Crown Jewels, on display close by in the Tower of London.

Gold is still a refuge in times of trouble. When currencies crash, gold gains in value. The post-9/11 political uncertainty is regarded as the main reason why gold prices have soared since. Wars are good for gold miners. Dictators from Hitler and Stalin to Ferdinand Marcos and his wife lusted after and obsessively hoarded gold. Gold made apartheid. And the biggest ever stock market scam. In the 1990s, there was feverish trade in a company that claimed to have found the world's largest gold mine, in the heart of the Borneo rain forest. Billions of dollars went down the drain when the find was exposed as a fraud.

But gold is also loved by the poor. More than half of all the gold ever mined is in the form of jewelry worn by hundreds of millions of people across the planet. Every Indian bride takes gold for her dowry. There is more gold stored in simple chests in tens of millions of Indian homes than in Fort Knox and the vaults of London put together. African peasants fleeing from wars carry gold as their only possession of value. The world's population agrees on very few things. The value of gold is one of them.

And I have my wedding ring. I can remove it from my finger, if I try very hard and apply a little Vaseline. But it feels like a betrayal even to do it, and I hastily put it back on—such is the power of gold.

At last we approach the gold face. The roar of three drills pounding into solid rock becomes so loud that even shouting into the ear of the person next to me is useless for communication. I look for signs of gold. Amazingly, here in the heart of the world's most productive mine, there is no sign. Not the slightest glimmer. Mel shows me the rock containing the gold. It is a dull gray quartz, with large pebbles embedded in it. These were the pebbles in a river delta 3 billion years ago, before anything more sophisticated than microbes lived on the earth. Before there was oxygen in the atmosphere, there was gold in these hills. Back then, the gold brought down by rivers from the mountains concentrated around the pebbles. Occasionally, if you look carefully, you can see a sparkle around the edge of the pebbles, he says. But miners can go years here, drilling in these black holes, without once seeing any indication at all of what they are devoting their lives to winning from the earth.

Down shaft 7 in Driefontein, I feel as if I am penetrating not just into the earth's crust, but into the psyche of humanity. A lust for gold may have been one of the first things that distinguished *Homo sapiens* from our fellow hominids. My footprints down this tunnel feel like the footprints of my species, tracing a search for something so elusive but so desired; something obsessional but sacramental; something venal but pure.

This journey into Driefontein is the start of my journey to find my footprint. This is in part a geographical journey to discover where my stuff comes from. But it is more. It is a journey of the psyche too. And gold sums it up. The pursuit of gold is in one sense the most absurd human endeavor. Environmentally and in humanitarian terms, it makes little sense. You might say it encapsulates the folly of mankind, and how we abuse each other and the planet. Like King Midas, we are cursed by it. And yet who would forgo our love of beauty, and of a metal that seems to touch us to our core? Of all the metals mined by man, none has survived so well, been recycled so often, or been cosseted as carefully as gold. It symbolizes some of the best in us too.

This is not a book about gold. But gold was a lodestone for my journey. And I wore my ring every inch of the way.

Part Two

My Food

Coffee 3
Throwing a Hand Grenade
into the Cozy World of Fair Trade

I think we were a bit pleased with ourselves, John and I, as we sat in a shed on the slopes of Mount Kilimanjaro. A sudden rainstorm had erupted outside, and water was crashing onto the corrugated-iron roof. But we were warm and dry, self-satisfied missionaries for the fair-trade cause. John Weaver—tall, diffident, but with a slightly military bearing—is the chief buyer of coffee beans sold in supermarkets by the fast-growing British fair-trade company Cafédirect. Taking the same trail as the buyers from Equal Exchange, a U.S. purchaser of fair-trade coffee from Kilimanjaro, he was on his annual tour to check the quality of the new coffee harvest, and to decide whose beans would go into the next year's packets of the company's top-selling Kilimanjaro brand.

We were visiting the Mwikamsae Kenyamvou Cooperative Society, whose beans had been the biggest component of the previous year's batch. John wanted to congratulate them on their beans. I too had a story to tell. I bought Kilimanjaro coffee in my local grocery, I told the twenty or so farmers assembled on benches in the shed. I had chosen their beans for years, because I liked them best. Now I had come to find out for myself who grew them. The farmers had come to puncture our complacency.

The meeting began formally. There was a brief prayer and discussion of the minutes. When John spoke, the farmers applauded as he announced that last season Cafédirect had bought 222 sacks of coffee, weighing close to 15 tons, from their nine hundred smallholder farmers. They listened intently as he described how the international price of coffee was determined each day in New York, and how this was relayed to the Tanzanian auction in Moshi, the town at the bottom of the mountain where all their coffee was sold.

There were rumbles of agreement when he acknowledged that low prices caused problems for farmers. And nods all around as he explained how the purpose of Cafédirect was to ensure them a reasonable return for their work. "We guarantee you a minimum price," he explained, "whatever the international market price might be. For your high grade of gourmet Arabica coffee that is one dollar forty-six cents a pound—twenty cents higher than the world market price at the moment."

John finished. I thought it had gone down well. Then Jacob Rumisha Mgase, an old farmer with a weathered face, got to his feet. "I speak for the farmers," he said. There was a hush. "We'd like to know how much our coffee costs in the shops in England." It was a hand grenade thrown into the cozy world of ethical trade.

John picked up one of his packets: "This pack of Kilimanjaro coffee weighs half a pound and sells for about three pounds. So a pound of coffee costs just over six pounds, or about twelve dollars."

His audience may have been peasant farmers, but they knew how to add. Jacob turned his hat slowly in his hand. "So you buy our coffee for one dollar forty-six. And sell it for twelve dollars. Is that fair trade?" His family, he said, could live for a week on the price of a pound of coffee in London.

John did his best. He explained that after the coffee beans leave the farmers, they have to be cured and graded and packed and shipped to Europe, where they are roasted, ground, packaged, distributed to retailers, advertised, and finally sold to the public. And even in fair trade everyone had to make a profit. To general bemusement, he pointed out the importance to supermarket sales of the smart metallized-polymer packet, made in Spain. Jacob responded that his farmers had to fertilize the soil, plant the coffee, grow it, pick it, depulp it, dry it, carry it to the co-op, weigh it, and put it into sacks. "All our costs are going up, but the coffee price keeps going down. We farmers are totally exhausted by what we do. I request on behalf of the farmers, try to think of us. Please pay more."

At the back of the room, as Jacob talked, a man wet from the rain came into the shed with a small sack and put it on the scales. It contained just 7 pounds of beans—the output of his smallholding for a month. His income for that sack was about $10. Not for low-grade beans destined for a jar of instant, but for acknowledged high-quality "gourmet" Arabica coffee, labeled as fair trade and destined for sale at a premium price to some of the most discerning coffee drinkers in the world. And me.

No wonder that these farmers, though educated themselves, struggle to send their own children to school. No wonder their roofs leak. No wonder that no farmer I met had even a motorbike to take his beans to the shed. No wonder they can't afford to replant bushes that were put in the soil by their grandfathers and are losing productivity.

Mount Kilimanjaro is prime coffee-growing land. The climate here, over 6,500 feet up the slopes of Africa's highest mountain and away from the heat of the equatorial plains, is perfect. The sandy soils are fertile, and the vegetation lush. Except during droughts, there is sufficient water. And the farmers are up to the job. In among the coffee, they keep a cow or two. They grow maize for their bread, finger millet to make a potent banana beer called mbege, and bananas for sale in the local market and to shade their coffee bushes.

After the meeting broke up, Jackson Kombe, one of the co-op's leading lights, showed me his house and 5-acre farm. He had 1,400 coffee trees that produced more than 660 pounds of beans in an average year. That was an income of about $1,000. His coffee was organic, because he didn't have the cash for expensive pesticides. His four cows provided organic fertilizer. He showed me a weed by the path: "I mix its leaves with water; it makes a bio-insecticide." Ecologists say this mixed, largely organic farming system is ideal for conserving water, soils, and wildlife. The farmers were doing everything right, and producing a superb product. Their problem was that their plots were tiny. And with coffee prices so low for so long, it has been hard for them to make a living.

Back at the shed, a truck had arrived from the Kilimanjaro Native Cooperative Union (KNCU), the "mother co-op" with which all the local co-ops around the mountain are affiliated. The driver was loading up with sacks of beans. Only about a fifth of them would end up with Cafédirect. The rest would be sold on the open market. They might end up in a high-priced cappuccino in Rome, a fierce black brew in Riga, or a Starbucks almost anywhere on the planet.

Starbucks. Now that word set off a hubbub in the co-op shed. Much as Cafédirect courts these farmers (their photos appear on Kilimanjaro packets), many barely recognized its name. But they knew about Starbucks. As the rain set in again, Jacob almost spat its name across the shed. They knew all about its thirteen thousand branches purchasing one in every fifty coffee beans around the world. They knew the price of a latte better than I did—$5 in some places, someone said. They knew that you could get more than sixty cups out of a pound of coffee. And they were quite capable of working out that this meant Starbucks charged $300 for coffee for which the farmers received less than $1.50.

That's a bit unfair to Starbucks. The company has branches to run and ambience to create. But everyone likes an ogre. And after thinking about Starbucks, the farmers liked John and Cafédirect rather more. One thing still rankled, however. Cafédirect was selling their coffee as fairly traded, and they just did not regard even its premium prices as fair. I was beginning to agree with them. John and I traveled back down the mountain, catching glimpses of its melting ice cap. This is the road to Moshi taken by Kilimanjaro coffee from some sixty thousand smallholders in almost a hundred local co-ops, where KNCU collects, cures, and ships beans. KNCU is the oldest cooperative in Africa, about to celebrate its seventy-fifth anniversary. It was set up in the 1930s, when the British colonial authorities were introducing coffee on the mountain and the local farmers felt they were being cheated on price. Its members produce about 5,500 tons of beans a year. In 2006, about 90 tons went to Cafédirect, in five shipping containers.

Moshi is Tanzania's coffee capital. Every bean of Tanzanian coffee is sold at the auction here. KNCU coffee is auctioned beside lots from smallholders' co-ops and large private coffee estates from across the country. The auction itself—held at 10 a.m. every Thursday morning—was a bit of a disappointment. There were no loud men shouting bids. Instead, the buyers sat at desks and each lot was displayed on computer screens in front of them. They pressed buttons to bid. Even the sound of the auctioneer's hammer was made by pressing a button. As ersatz as any instant blend.

Lots were sold, at around $1.25 a pound, to young traders who had flown in from Nairobi for the day. Their nametags bore the logos of local-sounding organizations like Mwanze and Mazao and Kilimanjaro Planting. But in reality, they were mostly the employees of global trading conglomerates. They included Volcafe Holdings, Ecom, and Edmund Schluter, all based in Switzerland; Neumann Kaffee from Hamburg; and the U.S.-based Louis Dreyfus Group. These traders of the most valuable commodity in world trade after oil sell to five big coffee manufacturers: Nestlé (the world's largest coffee titans and makers of Nescafé), Kraft (makers of Maxwell House), Sara Lee (makers of Douwe Egberts), Procter & Gamble (makers of Millstone), and the German giant Tchibo.

But I was most interested in Lot 198, the KNCU coffee chosen by John and destined for Cafédirect. Legally, it had to go through the auction, the same as any other coffee. But because Cafédirect pays premium prices, it normally gets what it wants. Sure enough, Lot 198, one container load of prime KNCU Arabica coffee, was sold to Twin Trading, Cafédirect's buying partner, for $1.46 a pound. It was the highest-priced sale of the day, and meant extra money in the pockets of the KNCU's farmers. Fair trade in action. Except that down on the farm, things look far from fair. Why so?

I don't think the answer lies with the motives of the fair traders. Their hearts are in the right place. In Britain, the fair-trade movement began as an initiative of the Greater London Council in the 1980s. The idea was to link cooperative stores in London to producer groups in the developing world, and cut out the middlemen. Nicaraguan cigar makers and Cuban textile workers were among the first beneficiaries. The concept didn't get going in the United States until the 1990s, when Transfair USA began certifying fair-trade products. From the beginning, the fair trade movement has been about more than the prices paid to individual farmers. It has also been about promoting co-ops and helping communities by providing extra premiums to approved community projects. On Kilimanjaro, Cafédirect hands back more than $40,000 in this way each year. The farmers I met were spending their dividend to set up a tree nursery and on scholarships so their children could go to secondary school. Money from other fair-trade buyers was also going into tourism, roads, and other local infrastructure.

That near-socialist philosophy persists as the movement has grown into a fully fledged capitalist enterprise, embracing fruit, herbs, spices, rice, yogurt, honey, flowers, and wine, and as the logo has moved from charity and church shops to mainstream stores. Cafédirect alone fills a billion British coffee mugs a year. U.S. sales of fair-trade coffee have exceeded 74 million pounds in the last six years, according to Transfair, sufficient to fill more than 4 billion mugs. Even the biggest coffee firms like Nestlé have their own fair-trade brands.

But fair-trade coffee is far from taking over the world. If Nestlé were a true convert, it would be paying a premium price for all its coffee rather than, at last count, just 0.2 percent. And therein lies the problem. Fair trade pays a small premium, but on a global coffee price that remains catastrophically unfair to the coffee farmers. After the collapse in 1989 of the International Coffee Agreement —a cartel that had managed production, kept prices fairly high and stable, and ensured that small coffee farmers had markets for their products—prices crashed, reaching rock bottom in 2001. By 2007, coffee prices were starting to creep back up, but they remained well below their level in the mid-1990s. And small farmers like the members of the KNCU have lost out most dramatically. While Cafédirect's sales of KNCU coffee may be booming, KNCU's overall output is only a quarter of what it was a generation ago.

I saw the decline most obviously on the last morning of my trip, when I visited KNCU's curing plant in Moshi. Curing is a fairly simple process of cleaning and hulling the beans and sorting them into different grades for dispatch to roasters. I expected a fairly modest factory, but instead I found a huge building almost twice the length of a (U.S.) football field and six stories high. In the 1920s, when it was built, more than seven hundred people worked there. Now the place was largely empty, with a part-time roster of fewer than seventy staff. "We have the capacity to handle a million sacks of beans in a year in here, but we only had eighty thousand sacks last year," said my guide. Beneath piles of dust, I made out the chipped and discolored remains of beautiful floor tiles. The place must once have been magnificent. But now it is a mausoleum commemorating the old state-run industry.

From Moshi, my Kilimanjaro coffee takes the potholed road to Dar port, where it is loaded onto a container ship for the journey up the east coast of Africa, through the Red Sea and the Suez Canal, then through the Mediterranean and around Spain to Tilbury dock on the Thames estuary and to a bonded warehouse close by. About twice a month, 3-ton batches leave the warehouse for a Dutch-owned coffee roaster called Gala in Dartford in Kent.

Gala is Europe's largest roaster of own-brand coffee. It has helped nurture Cafédirect's growth, to the point where it has become Britain's sixth-biggest coffee brand, selling more than 1,300 tons a year, including Kilimanjaro, Palenque from Mexico, and Machu Picchu from Peru. In the tasting room, Gala's technical manager, Nick Boxall, said he'd seen all the coffee fads come and go, but admitted that fair trade has given a new buzz to the industry. Was this more than just a new way for traders to make more money out of coffee? He smiled rather enigmatically.

Some see fair-trade coffee as cynical marketing of a "premium" product, an ethical veneer for an industry built on exploitation. That would have been grossly unfair back in 2001, when for a while Cafédirect was paying three times the market price. But today, the virtue is perhaps less clear-cut. Yes, the Kilimanjaro farmers get a small premium—an extra 10 cents from the price of every half-pound bag I buy. But most of the extra I currently pay in the supermarket is taken up by the additional costs involved in grading and inspecting and trading in the smaller volumes of coffee. Probably this is inevitable, but it does leave a bad taste. And it helps explain why on Kilimanjaro "fair trade" is merely slowing decline.

The Chagga people, who make up most of the coffee farmers on Kilimanjaro, are not surprised by this. They have a long and fairly shrewd acquaintanceship with global markets. This began in earnest more than a century ago, in the 1890s, when a Chagga king called Rindi established Moshi as a world center for the ivory trade. Where today the town's warehouses are full of sacks of coffee, back then they were lined with elephant tusks culled from animals killed on the plains west of the mountain. Most of the buyers were German settlers, who shipped the ivory out via Zanzibar to make billiard balls and ivory statuettes and piano keys for the drawing rooms of Europe. Quite possibly, the ivory keys on my piano back home were traded through Moshi. Doubtless, the Chagga were complaining about the price then, too.

Eventually the elephant population collapsed, and when the British took over in 1922, they introduced coffee. But the Germans are still around in Moshi. One set up the Salzburg Bar, where I ate. And a VW dealership still dominates the local automotive market. I was driven back to the airport in an old VW van. As we drove, I felt I wanted to shake the fair traders and demand that they charge me more. Jacob's accusing words came back to me: "Is that fair trade?" Why, I thought, should feeling virtuous come so cheap when it still leaves farmers so poor?

I did not fall out of love with the idea of fair trade. Far from it. Companies like Cafédirect are on the side of the angels, fighting the global giants who constantly attempt to drive down prices. I drink fair-trade coffee, and would encourage you to do the same. But I do think it is a misnomer. Fair-trade coffee is not fairly traded. The price is still dictated by market conditions in Britain and America, rather than living conditions in Tanzania. But the fault is with us, the consumers, not the people of Cafédirect. The critical question is how much extra we consumers are prepared to pay, as we peruse the coffee packets on display in the supermarket, in order to feel good about our coffee. So far, we are not prepared to pay very much. We want our ethics on the cheap. If we convince ourselves that we are paying a fair price, giving the coffee farmers a proper return, then we are deluding ourselves.

Wild Things   4
The Last Roundup on the
High Seas and in the Hills

I love fish. More than meat, really. In Britain, that means haddock and cod, mackerel and Dover sole, crabs and oysters—all from the seas around our coasts. Other coastlines have other species. But the terrifying disappearance of marine stocks suggests that ours could be the last generation to enjoy this pleasure.

Everybody in England knows that the North Sea cod is on its last fins. Grimsby on the east coast of England, once among the world's largest fishing ports, still has the world's biggest fresh-fish auction, but only 1 percent of the sales are from domestic boats. And the North Sea is only following the disaster already visited on Newfoundland and the Gulf of Maine, where New England fisheries have collapsed in recent years. Cape Cod was once worthy of its name, but no more. And this is a global crisis. Scientists say that most of the world's commercial fisheries will be gone by midcentury. Soon there will only be farmed fish.

What place, I wondered, would illustrate this crisis best? I have been appalled at the squalor of Asian fish factories in Kisumu, the Kenyan port on the shores of Lake Victoria. And at seeing Filipinos using squeeze bottles full of cyanide to stun reef fish so they can be caught live and sold days later in the restaurants of Hong Kong. I have watched mussels being harvested from the lagoons of Venice and eaten Atlantic tuna in Tenerife; sat cross-legged to consume delicious smoked cod in Kyoto and picked at unnamed fresh fish straight from the waters of the River Taz in Siberia, the Ganges in Bangladesh, the Amu Darya in Uzbekistan, and an ice hole near Yellowknife in Canada. But I remember most a journey to find out why

Europeans like me can eat fresh fish from a strange desert state called Mauritania.

As the sun set over the Atlantic, Italie de Silva worked fast: gutting the sharks, rays, and guitar fish, cutting off their fins, and laying out the flesh to be salted by the incoming tide. It was a good catch, he said. Good for recent times, anyway. Two days at sea had yielded maybe a hundred fish, worth $3,000. As he slit the bellies of the female sharks, fetuses spilled out. He threw them aside to be eaten by birds. He didn't seem too concerned. But those fetuses should have been next year's catch. The fishermen were, in effect, killing many fish for the meat and fins of just one.

This scene took place on a beach by the tiny village of Iwik in the Banc d'Arguin, a giant national park in the isolated West African state of Mauritania. The country is virtually all desert, but the park's coastal fringe contains one of the world's richest fishing grounds. For now. Because the fate of the shark fetuses sums up an escalating crisis here. A crisis in which hundreds of European trawlers that bring shark and squid, lobster and bream, hake and grouper, mullet and much else to the restaurant tables of Europe are guilty—hook, line, and sinker.

Fish from the Banc d'Arguin feature regularly in British fish markets. You can dine out on Mauritania's finest natural resource in the restaurants of south London. But soon our plates will be empty. Already some species are gone. "In the old days, we could see the mullet coming," remembered Mohammed ould Swidi, Iwik's village chief. "We just walked into the water with our nets to catch them." But that was before the rest of the world got wise to the riches here.

The Banc d'Arguin is a large submerged sandbank extending from the Sahara out into the Atlantic. Cold waters well up from the ocean depths near here, bringing to the surface a rich reservoir of nutrients. As the nutrients flush across the Banc, they feed plankton and sea grass. These rich marine meadows are the prime breeding and nursery area for a fishery that stretches more than 1,200 miles along the West African coast from Morocco to Guinea-Bissau. But as fishing grows more intense, the fishermen are drawn ever closer to the source of this marine fecundity, the Banc itself. And that threatens the survival of the entire ecosystem.

Thirty years ago, European conservationists, led by the Swiss industrialist and cofounder of the World Wildlife Fund (WWF), Luc Hoffman, went to the Banc and persuaded the Mauritanian government to declare it a national park. The conservationists were mainly interested in the vast birdlife that feeds on the fish there. In winter, the Banc has the largest collection of wading birds in the world. But protecting the birds protects the fish too. Or it did.

The only fishers allowed in the park are the Imraguen, the poor black desert inhabitants alternatively ostracized and enslaved by their Arab and Berber masters. And under the park's rules, they have to use traditional sailing boats. That system would have worked well except for events outside the park. For on the fringes of the Banc, a fishing free-for-all has taken hold.

First, thousands of African fishermen, in motorized versions of their traditional boats, called pirogues, started to make the journey north from Senegal, Gambia, and even farther afield. They cruised the waters outside the park, taking rich pickings but leaving the nursery intact. Then came the European trawlers, sold licenses by the Mauritanian government to "fish the line" 7 miles from the shore that delineates the edge of the park. It was after the trawlers came, say the Imraguen, that the mullet stocks collapsed. Since then, the octopus catch has been halved and sawfish have disappeared completely, along with hammerhead and tiger sharks.

Theoretically, the trawlers and pirogues can share the waters outside the park. But in practice, it is far too dangerous for the small pirogues to jostle with factory ships that can be close to 500 feet long. Collisions are frequent, as are deaths. So pirogue captains head inshore, into the park, where rangers regularly impound their boats for illegal incursions.

While visiting a tiny park fishing village on the site of an old Portuguese fort on Agadir island, I met Lamin, a nineteen-year-old poacher. He had no money and little food, and the only bed he could call his own was 600 miles away at his mother's house in Gambia. His boat had been impounded here for two weeks. His lodging, until the owner of the boat showed up to pay the fine, was a tiny asbestos-roofed hut on the beach that he shared with twenty other captive poachers.

Lamin told me that he and his fellow crewmen—two Senegalese and a Mauritanian—went to sea for a week at a time in the open 25-foot boat. Its owner paid them about $80 per trip. "I need the money for my mother, who is ill," he said. Dozens of pirogue boats like his lay their nets in the park at night when the patrol boats have little chance of catching them, and take off before dawn. He got caught because his navigation went wrong.

As the world's other great fish stocks are decimated, more and more fishermen like Lamin come to the Banc, and more and more people come to rely on their catches. Farmers from the arid lands of the Sahel are turning to fishing to provide protein for their families. Hann beach on the edge of the Senegalese capital Dakar has seen a huge boom in recent years. Bira Gueye told me that when he first came, in the late 1960s, "there were just five boats here." It was hard to believe as we walked down the beach past the Yamaha outboard motor showroom and the large freezer warehouses. There are now more than a thousand boats regularly pulled up on the sand.

But here too, the fish are disappearing. We watched several crews head out to sea. In the old days, said Bira, he could sail for twenty minutes and fill his nets. "We were so close as we drew nets that we could see people having lunch on the beach. Now it takes four hours and eighty gallons of petrol just to get to the fish." Many crews go out for two weeks at a time. They take ice boxes with them to keep the fish from rotting before they get them to shore.

On shore, they sell to traders on the beach. I met Victoria from Ghana, who buys from beach traders and regularly ships home 40-foot containers full of salted shark meat. She makes enough, she told me, to have one child training to be an airline pilot and another at an American university. Her grandson was beside her, learning the trade. But she wondered if there would be any fish left by the time he was grown up. Also meeting the boats as they came ashore was Dimas Santos, a Portuguese fish exporter. He bought fish such as grouper,

bream, and hake from the pirogues for sale at high prices to European restaurants. As we toured his packing plant, he told me he used
to buy more than 30 tons of fish a day from local fishermen, but now
it was down to around 10 tons. "There just aren't the fish in the sea
anymore," he said. "We are paying the price for years of overfishing."

The pirogues are routinely blamed for the collapse of the Banc
fisheries, partly because only their activities fall foul of the law. But
here it is the law that is at fault. For the pirogues, though huge in
number, take only about a twentieth as much fish as the trawlers out
on the horizon.

The European Union has for a decade now bought the rights to
take most of the fish from these waters. In 2006, it signed a new
deal with Mauritania worth 86 million euros a year, allowing access
for two hundred boats to catch some 55,000 tons. The EU's fisheries
policy states that all fishing by European trawlers should be "sustainable." But that brings a hollow laugh on the beaches, as well as
from industry observers. "Foreign trawlers are strip-mining African
waters. It is a scandal," says Callum Roberts, a marine biologist from
York University. "They are wrecking the future of African fisheries."

The trawlers mostly unload either in the Canary Islands or back
in Europe, so the nearest I got to them was watching forty slow-
moving white blobs on the radar screen at the Mauritanian navy's
coastal tracking station in the park; each blob was a trawler from
Spain or the Netherlands, China or Portugal. "It's like a town out
there," said Antonio Araujo, the Portuguese park manager. "They are
taking everything," he said.

It is not just Mauritania. Foreign trawlers take almost as many
fish from neighboring Senegalese waters. According to a report from
the United Nations Environment Programme, the catch has had "a
serious impact on local food supplies." But the West African governments feel powerless. Fish make up two-thirds of Senegal's export
revenues. "These countries have large debts. They cannot refuse the
EU," said Pierre Campredon, a French marine biologist and longtime
adviser to the Mauritanian government.

What should be done? The WWF, which is one of the sponsors
of the park, says the trawlers should retreat and be replaced by a small,

élite trade using line-fishing boats and selling to high-price traders like Dimas, and by high-rolling tourists. Such a strategy might mean the governments could continue to harvest cash from Mauritanian waters, and it might be environmentally sustainable. But why should the upmarket sellers to European restaurants have priority over the pirogues? Surely, they have first rights to the fish. The way the locals see it, Europeans have taken over their waters. One group keeps them out of the park to protect the birds; the other group keeps them out of the ocean so the fish can fill bellies in Europe.

I never met the European trawlermen. But I did meet the "conservationists." On my last night in the park, a group of overweight European businessmen turned up. They had come to watch birds. It was their perk for helping fund the park. But they seemed like a hunting party from another age, lounging in luxury tents erected by local laborers. They had no interest at all in the people living in or dependent on the park. They just came to see birds. I don't like that kind of environmentalism.

Meanwhile, as the fish disappear, the pirogue owners have stumbled on a new business—smuggling destitute Africans, many of them fishermen no longer able to make a living, to the Canary Islands. The bigger boats can cram a hundred people in. If the tides are wrong, the boats deliver emaciated and sometimes dead would-be migrants onto the beaches of Tenerife and Fuerteventura. But the number of people coming to the beaches and willing to pay for a ride into the unknown continues to grow. Granting their wish, the boat owners say, is more lucrative now than fishing. If the developed world wants the fish, some might say, it has a duty to take the impoverished migrants too.

We still hunt for fish. But otherwise, there are not so many foods that we gather from nature. Outside hunting estates, the chase is a distant memory. Our meat is domesticated—and frequently industrialized. But we do still consume food from the wild. I was surprised to discover that one wild delicacy in many kitchens is oregano, my favorite herb. I have a small jar in the kitchen, and when I am cooking I add it more often than any other herb. A lot of the world agrees with me. Its slightly bitter, vaguely minty flavor is the signature of Italian pizzas and pasta dishes. The word *oregano* comes from the Greek *oros ganos,* or joy of the mountain, and most of it comes direct from a mountainside.

There are lots of local varieties of oregano growing wild around the Mediterranean. In my local grocery store you can buy fresh oregano from Israel (and sometimes next to it are packs whose labels say they come from the West Bank, though I wonder if it is the same stuff labeled for customers of different political persuasions). But most is from Turkey, which exports about 7,700 tons of dried oregano leaves a year—a figure that has doubled in less than a decade. Someone worked out that this was enough for 7 billion slices of pizza, so everyone on the planet could get a piece.

Every Turkish hillside is said to produce oregano with a different taste, but the best is hand-cut by peasants with serrated knives in the spectacular limestone hills of the Sütçüler district in Isparta in the west of the country. Each summer, whole villages leave their homes to scour the gorges and hilltops, taking branches of the dwarf shrub back to their villages, where they dry the leaves and pack them for sale to international buyers. For a few weeks, the world comes to Sütçüler.

In theory, oregano should be easy enough to cultivate, but it is not done much because the resulting plants are inferior. There are different stories about whether all the harvesting is endangering the plant in the wild. Stefano Padulosi of Bioversity International (formerly the International Plant Genetic Resources Institute) in Rome, an oracle on obscure food crops, says, "Some of the varieties are disappearing before we know much about them. One indigenous variety in Crete has almost gone." He fears for the Turkish varieties, too. But interestingly, in the Sütçüler district, villages like Candir, Sarmemetler, Gumu, and Buydilli have responded to the soaring popularity of their ancient herb by establishing their own cooperatives, with strict conservation rules.

The Mediterranean harbors many wild herbs. In Spain, some 80 million thyme plants are uprooted every year for their leaves and oils.

Wild chicory is picked in Lebanon, and wild capers in Syria. Until the recent explosion of interest encouraged cultivation, the salad vegetable rocket was little more than a southern Italian weed with a half-forgotten culinary pedigree.

This set me thinking about what else I eat that comes direct from the wild. With the exception of fish, the options are limited. I have eaten wild rabbits and pheasant occasionally, and supposedly wild boar. I once nibbled at whale and seal meat on canapés in northern Norway, and turned down wild rat at a roadside chophouse in Ghana. I am told you can buy shrink-wrapped wild elephant in some African supermarkets. (Probably the biggest market is in Chinese medicines, and I will come back to that in a later chapter. The Chinese eat everything.)

I have never personally hunted. But I have frequently gathered blackberries from the hedgerows of England, and also sloes—the fruit of the blackthorn tree. Our household uses sloes to flavor gin, which I thoroughly recommend. I have picked dandelion leaves for a salad. Once in the northern Siberian town if Krasnoselkup, I had a whole meal of wild mushrooms and berries. And in bars in El Paso you can buy a spirit, rather like tequila, made from a wild cactus picked over the border in the Chihuahua desert.

On a visit to Turkey a few years ago, I came back with a packet of salep, a flour made from the boiled tubers of a local orchid. When heated with milk, the flour made a good evening drink, a bit like Ovaltine. Salep was popular in England before tea and coffee. The word is Arabic for fox's testicles—a description of the shape of the tuber, apparently. I discovered only after bringing the packet home that the orchid is mostly harvested from the wild and is now highly endangered. Two pounds of flour require a thousand orchids, so my small brown box bought at a supermarket in Istanbul had uprooted at least two hundred of them. I then found that the export of salep from Turkey is banned. Probably I could have been clapped in a Turkish jail for smuggling it out. I kept thinking of the film *Midnight Express* whenever I looked at the packet, and I eventually threw it out.

The sage in my kitchen might be harvested from the wild. The jar just says it is a product of Turkey. But herb world insiders like Al

Goetze, who buys for the American company McCormick, say "Turkish" sage is usually picked from the wild in the mountains around Albanian towns like Tepelene, Berat, and Shkoder and shipped out via Montenegro and Istanbul. According to Goetze, "Albanian is the best sage in the world." Sage harvesting and export, he says, was about the only part of the trading economy that worked under the bizarre Communist regime of Enver Hoxha, in the days when Albania was a Cold War equivalent of North Korea today—so isolated it made enemies even of its ideological friends.

Albania continues to export half of all the world's dried sage and medicinal sage oils. But now the 2,700-ton-per-year business is in private hands, allowing people like Mehmet Guga from Tepelene, who spent half a lifetime running the business for Hoxha, to grow rich in their old age. Almost all the sage is harvested from the wild by hand using a sickle. The leaves are dried in the sun in the villages, rather like Turkish oregano. Certainly, when I stuff a free-range chicken with sage and onion, the stuffing is much wilder than the chicken.

Most herbs and spices have long since moved from wild harvesting to cultivation. But the hint of the exotic often lingers. The forested slopes of the Moluccas, or "spice islands," between the Philippines and New Guinea, were once the only home of the trees on which cloves and nutmeg grow. These spices have always been in huge demand in Europe. Cloves cure bad breath, and nutmeg preserved meat in the days before refrigeration. The casing surrounding the nutmeg seed provides mace.

When the Portuguese explorer Ferdinand Magellan embarked on the first round-the-world cruise in the 1520s, the bill was paid by selling Moluccan cloves. And, after the Dutch captured the islands, their spices were worth more than gold on the wharfs of Amsterdam. The Dutch prospered for centuries on their monopoly over these spices, till a French diplomat stole samples and established plantations in the Indian Ocean. The descendants of the pilfered cloves eventually turned Zanzibar into the largest producer of cloves, a position the small island holds to this day. If you are ever there, I recommend buying the clove massage oil, too.

Similar stories are told about the taming of cinnamon (from Sri Lanka), ginger (from southern China), and cardamom (from India), though the crocus that produces saffron, which was once the most expensive of all spices, is a sterile mutant unknown in the wild.

I enjoy all these herbs and spices. But most of all I love vanilla. When everyone else is buying exotic-flavored ice creams, I often stick to its familiar and subtle taste. Vanilla is a bean from Latin America that grows in long pods on a climbing orchid. But today, two-thirds of the world's crop is cultivated on small mountain farms beneath the rain forest canopies of Madagascar and nearby on Réunion and the Comoros islands in the Indian Ocean. There it is called Bourbon Vanilla after the French kings who transplanted the orchid two hundred years ago. Cultivating the bean in Africa is not easy. In its American heartlands, vanilla is pollinated by a small bee. But without the bee, African farmers must tour their fields each morning looking for newly opened flowers, which they pollinate by hand that day, before the flower wilts and drops off.

After picking, the ripened pods are boiled, and then cured by alternately leaving them to sweat in wooden boxes and exposing them to the sun—a laborious process that lasts for several months. The end result, as Goetze puts it, is "a rich, dark brown, moist and pliable bean that is loaded with aroma and flavor." Then villagers sell the pods to agents of the world's biggest vanilla trader, AGK, who take the beans to the coast for export around the world.

Vanilla has gone from boom to bust in recent years. Cyclones in the Indian Ocean destroyed much of the crop early this decade. Prices soared; farmers replanted to cash in. Uganda and Kerala, the spice capital of India, tried to join in the bonanza. But when the new crops matured in 2005, there was a glut. Production far exceeded the world demand for 1,000 tons a year. Prices crashed again, and a year on, in late 2006, they were still only at 5 percent of their 2003 level. The crash has been a disaster in particular for Comoros, where vanilla is the main employer and export crop.

Many herbs, spices, and natural medicines have a long history. But some have risen from total obscurity to become global brands only in recent years. One of my favorite stories is the rise of tea tree oil, a natural antiseptic frequently used in our house. Until the 1970s, the tea tree was known only to a handful of traditional bush pharmacists in the Australian outback. Then a young Australian hippie called Christopher Dean got a nasty infection under his toenail while traveling in Africa. It would not heal until, upon returning home, in despair he tried a bush "healing oil" made from the leaves of the tea tree. It worked—within minutes.

The tea tree had become rare as the open land of northern New South Wales fell under cultivation. And most trees produced only poor-quality oil. So Chris and his friends began a search for surviving stands that could produce quality oil. They hit pay dirt with a single grove in the remote Bungawalbyn basin. Like other old hippies who have turned into astute capitalists (from Steve Jobs to Ben and Jerry), soon a motley crew had set up camp to distill the oil. Local authorities repeatedly tried to evict them as undesirables, but the entrepreneurs eventually established a small plantation and began selling the oil. Fellow hippies bought it first. Gradually, word spread about its ability to cure everything from pimples to stings and burns to vaginal infections. Today tens of millions of trees have been planted to meet soaring demand, reversing a two-hundred-year decline for the "healing tree." Take it from an old hippie: it works.

I have a no doubt romantic liking for things from the wild. But we have to be realistic. Where demand for products is high, the chances are that the resource, whether animal or vegetable, will be hunted to extinction. In general we have to accept that domestication is essential to feed a world whose population is approaching 7 billion people. Especially if we want some nature to survive. But what Jack London described as "the call of the wild" is in us all. And sometimes, even if it is just going blackberrying on a sun-soaked autumn afternoon, I hope we can still respond to it.

Curried Crustaceans   5
The Weird World of Mr. Prawn

The curry houses of Britain are coy. I have often wondered where the prawns (the usual English term for shrimp) in my Saturday-night curry come from, but I have never gotten a straight answer. While the supermarkets are keen on "traceability," and refuse to sell prawns if they don't know where they are grown and how, curry houses don't go beyond the rhetoric of their own menus. I have always assumed that, because most British "Indian" restaurants are run by Bangladeshis, Bangladesh is where their warm-water king prawns come from. My hunch proved right.

And there is, it turns out, a clear leader in the business. His name is Iqbal Ahmed, or "Mr. Prawn," as he is known, after one of his brands. He is a big fish in the British Bangladeshi community, and a hero among restaurant owners. I went to see him at his headquarters in the great English city of Manchester, right next to the Manchester City soccer stadium. In the boardroom, surrounded by photographs of him shaking hands with royals and political leaders, and collecting his Order of the British Empire from Prince Charles, he fed me plates of breaded prawns and the story of his success.

Iqbal is the son of Bangladeshi parents who fell on hard times and moved to Britain. "Our family were landlords, but we lost our land," he said. In Britain, his father ran a small grocery shop in nearby Oldham, "open all hours." Iqbal has restored family fortunes by creating a $400 million prawn business, called Seamark, which has two processing plants, one in Chittagong, the main port in Bangladesh, and the other in Manchester. His big coup came in the 1980s, when he introduced Britain to black tiger prawns. They spread fast, making him one of the twenty richest Britons of Asian origin today. And they have allowed him to expand his empire around the world, including a large cold store in Brooklyn that sells Bangladeshi prawns across North America.

Mr. Prawn is now diversifying. Back in Manchester, Mr. Prawn's cash-and-carry operation sells vegetables, meat, and anything else a curry house could want. The day I visited there was a container of frozen chicken from Brazil sitting outside. But prawns remain the main business. He shifts a million of the beasts a day. Half of all the king prawns sold in British restaurants—Chinese and Thai as well as Indian—are bought from Seamark. Exports are now 70 percent of sales, making Manchester the unlikely prawn capital of Europe.

As I worked through a second plate of prawns, their proud purveyor told me that my Saturday night curry most likely came from Chittagong in freezer containers through the Suez Canal to Felixstowe or Southampton docks, before taking a truck to Manchester. But when we got down to the nitty-gritty of how the raw prawns got to his processing plant in Chittagong, Mr. Prawn told me that he buys from brokers and never goes near the farming end of things. So I figured I would have to go and find out for myself.

On my journey across Bangladesh, I found the prawns that go into my curry, the farmers who grow them, and the men who supply them to the world market. It is a major business. It ships out 55,000 tons of prawns a year, mostly to Britain and the United States, and is the country's second biggest export earner. But its ecological footprint is huge. In pursuit of prawn pounds and dollars and euros, landowners have flooded some 800 square miles of the delta lands in the south of the country. And, far from spreading wealth, the trade is miring millions in misery, poverty, and corruption. I stumbled on mob bosses and their "musclemen," and cycles of debt and dependency that made me wonder whether I should boycott my Saturday tandoori.

I went first to southwestern Bangladesh, which has 80 percent of the country's prawn farms, including most of those growing black tiger prawns. I headed for Khulna, a city of 2 million people with virtually no cars, only a few motorbikes, and tens of thousands of cycle rickshaws. It must be one of the few cities of its size where human muscle power has yet to be replaced by fossil-fuel energy in transportation. What wealth there was on display was under the control of the new prawn oligarchs. Apart from flophouses, there were only three hotels in town—all owned by the princes of prawn.

Prawn farms stretch from Khulna to the Bay of Bengal. In the past three decades, they have largely replaced the old landscape of mangrove swamps interspersed with small farms. The swamps once protected large amounts of wildlife, including the famous Bengal tiger, and acted as a nursery for fisheries. Now the tigers are giving way to tiger prawns, which are not the same thing at all.

As the landscape has been transformed, a whole social system has disappeared, explained Ashraf-ul-Alam Tutu, head of the Khulna-based non-governmental organization (NGO) the Coastal Development Partnership, when we met in his Khulna offices around the corner from one of the prawn hotels. Once there were rice paddies and mangroves, with creeks full of fish, ducks in the yard, and cattle and birdlife and grazing pastures—much of it on commonly owned land available to the poorest as well as the richest. Now there is only private land and prawn ponds. The pastures are gone and the rivers are almost empty of fish.

Outside Khulna is a small administrative district called Dumuria. There are more than five hundred prawn farms here. I went to see Amal (not his real name), a strapping Bengali of about thirty whose father grew rice here and had chickens and a couple of cows. But now prawns are the main business. It was the start of the tiger prawn season, and Amal was throwing nets into his pond. I recognized the crustacean instantly from many a tandoori. He picked it carefully out of the net and threw it back to grow a little more.

Amal and his wife, a feisty woman whom he insisted was his "business partner" too, looked like a model for hard-pressed but determined business activity. Independent producers using local resources to create a product destined for a global market. Heroes, you might have said, of grassroots globalization. But it wasn't quite like that.

Amal's farming methods, like most out here on the delta, were simple. The only fertilizer for the ponds was dung from his two cows. Farming was organic by default. His one-acre pond produced just over 220 pounds of prawns in a year. This sounded like a pitiful quantity, only a tenth of typical yields in rival exporting countries like Thailand. But I learned later that it was actually slightly above the average for the Dumuria area. Amal said it brought him an annual profit of 33,000 taka, or $500 a year.

The scene was poor but tranquil, deceptively so. After we had been talking a while, Amal mentioned that he had problems with a big local landowner. About ten musclemen worked for the owner of the ponds between Amal and the river. They were threatening to stop the water from reaching Amal's pond unless he paid a hefty bribe. I tried to track down the culprit. Close by was a giant pond covering 75 acres. It was wider than the river that supplied it with water. As soon as we arrived there, we were besieged by a gang of workers. They told me the name of the owner. "He is a big political leader and a big muscleman," someone said. The pond was surrounded by huts where his workers lived, keeping watch on the pond day and night in case of prawn thieves, they said. Was their boss the man threatening Amal? Could I meet him? There seemed to be a wall of silence, and I was left to guess about that. "He doesn't live here" is all they said. Amal was lucky to still have his land. Millions of rice farmers have been bundled off theirs, often illegally, by big landowners anxious to maximize their prawn profits. A quiet terror has spread across the delta, with musclemen responsible for dozens of murders and rapes and hundreds of injuries. The killings are well documented. But the local police, who are clearly in cahoots with the gangs and their bosses, rarely act.

On local advice, I did not always admit to being a journalist. Around here investigative reporting is a dangerous profession. Seven months before my visit, a local reporter for a national paper was beaten up by thugs and then arrested, tortured, and held overnight by the Dumuria police, apparently for writing about land grabs. But he got off lightly. During the previous decade, fourteen local newsmen had been murdered in the area, including successive presidents of the Khulna Press Club. The NGO Transparency International calls Bangladesh the second most corrupt nation on Earth. I began to see why.

That evening, I met a group of landless women, a small delegation from the five thousand or so landless people round Dumuria alone. They had waited for hours to see me. One by one they stood up to tell their stories as we sat in the dark, with mosquitoes from the ponds besieging us. One woman told me simply, "We are poor. We have no shelter and no proper work. The musclemen have taken our land." They said the government had a policy of settling homeless people on unused state-owned land. But corruption was rife, and usually the big landowners grabbed the spare state-owned land.

Until recently, many of these women earned money by collecting prawn fry from the rivers. Fry are the basic raw material for any prawn pond. In the early days of the industry, there were enough in the rivers to seed the ponds naturally. But as the mangroves declined, so did the supply of natural fry, and a huge informal industry developed to collect them. Up to half a million of the country's poorest people in the remote delta regions spent their days up to their waists in water, netting fry from tidal creeks and rivers.

Inevitably, most of what they caught was the fry of other species, including fish. Most were simply dumped on the bank. Now the fish are disappearing too and the government has banned the harvesting of fry. Instead, farmers must buy from new commercial hatcheries owned by the industry's big players. The hatcheries turn out billions of fry a year and air-freight them across the country. This is clearly an environmental improvement. But few outsiders seem to have thought about the social impact of the ban. Hundreds of thousands of collectors have lost their livelihoods. There are, so far as I could gather, no statistics on what is happening to them. And few efforts to help them. Hence the women in front of me in the Dumuria night.

The new commercial fry hatcheries are mainly in the southeast of the country, adjacent to the coastal waters, where the brood stock—the "mother prawns"—are trawled from the open ocean. One mother prawn can provide millions of fry. But already the mother prawns are disappearing: another twist in the ecological downward spiral created by the prawn industry.

The impression I gained was of a delta ecosystem in meltdown, thanks to prawns. And yet the economic gains for all this destruction seemed so tiny. Amal's prawns might end up on my plate in London at almost $20 a portion. But his profits from that were a few cents. Why? His yields were poor, certainly. I could see that. But somewhere between south London and Dumuria most of my money disappeared. So I began to follow the trail of his prawns to find out where.

Amal packed his black tiger prawns in ice and sold them at the door to middlemen. The middlemen provided him with credit to buy his fry and maintain his business, so he had little choice, he said. Anyway, if he tried to bypass them, the musclemen would come around. The middlemen in turn sold Amal's prawns to gentlemen at a depot a mile away in Dumuria. When I dropped by the depot, the main activity seemed to be collecting the boxes brought in by the middlemen, adding more ice, and then selling them to a new set of middlemen equipped with vans to take the prawns some 6 miles to the wholesale market in Khulna.

So, back to Khulna. The wholesale market there was a long line of workshops down a street close to the ferry port. I watched one worker sorting a large pile of prawns by size, and putting them into different boxes. That, he said, was all they did. Oh, and add more ice. His boss said this one workshop dispatched about 330 tons of prawns a year to one of the city's numerous processing plants. He had a markup of about $200,000. There were about three hundred registered wholesalers here, with a throughput of more than 330,000 tons of prawns. With all these middlemen providing what might at best be called minimal services, I began to see why Amal had so little income.

Huddled close to the wholesale market were the processing plants. I visited three. They were spick-and-span, all stainless steel and tiled walls, refrigeration and running water. Lines of uniformed workers removed the heads and shells, sometimes cooked and always froze and packed the prawns ready for dispatch. In general, the hygiene rules seemed to be observed, though in one plant there was a brief panic when I lifted my camera in a packing area where none of the young women workers were wearing the required plastic gloves.

The biggest processor was M.M.A. Salam, the mustached owner of two large plants and vice chairman of the Bangladesh Frozen Food Exporters Association. His main brand was Castle Salam—also the name of his hotel. Iqbal in Manchester is his biggest customer. "He came here for our opening ceremony in 2001. We sell up to thirty containers a year to Seamark." That is more than 330 tons. With his own large processing plant in Chittagong and as a big buyer of prawns in Khulna, Iqbal seemed as big an exporter in Bangladesh as he was a seller in Manchester.

After visiting some traders selling prawn fry, I thought that at last I had seen the full extent of a ludicrously overelaborate supply chain, with far too many hands taking a share of my curry-house bill. I counted six sets of middlemen taking their cut. There were fry wholesalers and fry retailers; the guys who purchased prawns from the farmers; those who traded at the local depots; and those who sold to the Khulna wholesalers and the processors. But I felt there was a missing link. Then I met Dadu, who explained it all.

Dadu lived in a grand house in the center of Khulna. He was one of the pioneers of the prawn business here, a well-traveled man and a proud "freedom fighter." As a military mariner, he said, he had sent sailors on suicide missions to destroy Pakistani ships during Bangladesh's war for independence thirty-six years before. He clearly despaired about where his industry, indeed his country, had ended up. As his servant brought tea, he mentioned that I had missed a link in the chain, and it was the most important link. It turned out that, even on the 100-yard journey between the warehouses and the doors of the processors, more middlemen intervened. Dadu called them "the unseen," the moneylenders. "They operate behind the scenes at the Khulna warehouses. They are the real problem in this industry," he said.

In the absence of an efficient banking system in Bangladesh, "the unseen" effectively run the entire industry. They control the flow of prawns to the processors. And they provide the loans that trickle, at extortionate interest rates, all the way down the line to the farmers. The cash allows business to be transacted before Iqbal and the other processors and purchasers in Europe and North America pay up. Without them, Iqbal would have no prawns, and nor would my local curry house. Nobody would introduce me to "the unseen." And I found no academic who had ever done a study of their activities. They were just that: unseen.

Back in Manchester, I wondered again why Iqbal would not use the power of his company, Seamark, in the industry to cut out the middlemen, raise standards, and improve the lot of farmers like Amal. He had an interest in doing so, I thought. Bangladeshi prawns still have a poor reputation for quality and reliability. Western retailers say they pay 10 percent less for them than for others. It is self-evident that in Bangladesh this is a business mired in inefficiency, exploitation, and corruption.

Iqbal agreed about that. He had told me before my visit: "Things have not been done properly. There are few auctions and I can't deal with a thousand farmers all delivering two pounds of prawns in baskets. So I have to deal with the middlemen. I'd like to see a proper auction house in Khulna, but the local processors don't agree. They think I just want to buy everything."

What was needed, surely, was some kind of certification scheme, where prawns with a clear supply chain, ethical standards of production, and good quality could command higher prices. The "sustainable shrimp," in other words. And the Aquaculture Certification Council, a U.S.-based organization, had tried to set up just such a system in Bangladesh, with the backing of NGOs and aid agencies. It failed, according to Bill More of the Council, for want of interest among prawn processors. "Seamark, one of the largest exporters, has not shown any interest," More told me. "It could have made all the difference." Tutu, the social reformer in Khulna, was of a similar mind. "We want to use efforts by the corporate sector to set up a certification system to introduce rules about social responsibility. But it is hard to get the processors interested."

When I raised this with Iqbal, he was unapologetic. He knew

about the certification schemes being mooted, but wanted nothing to do with them. They were, he told me, designed to extract money from him. "I'd have to pay for certification, but what would I get in return?" Then he launched into an attack on the NGOs, whom he accused of trying to shut down the industry. "If the shrimp industry gets closed down, the entrepreneurs will be okay, they will go and do something else with their money. But the farmers will die of starvation if that happens," he said. "You'll punish them." Ouch. Why that sudden switch from "they" to "you"? I had the feeling his growing hostility to my line of questioning was now addressed at me personally.

I took his point that boycotts can be dangerous things, and that Bangladesh is a far from easy place to do business. Things will always be messy. However, there did seem to be an inconsistency between what he was saying to me and his presence just a couple of weeks before in the British government's "champions' group" of industrialists. Set up to promote "sustainable sourcing" in the food industry, it had the declared aim of doubling the amount of food in supermarkets covered by ethical trading schemes. But in his own industry, Iqbal seemed indifferent to the scheme on offer.

So should I keep on buying Bangladeshi prawns? In Khulna, Tutu said he had campaigned to keep out the prawn barons when they first came. But he now accepted they were there to stay. So he was working to improve conditions for the 4 million small prawn farmers and landless workers in the industry. "We want to improve the industry, not destroy it."

But in Dhaka, on my way home, I dropped in on Bangladesh's most outspoken campaigner for the poor: Khushi Kabir of an NGO called Nijera Kori. "I'm against the industry—period," she said. "If I could, I would shut it down." She doesn't buy the reformist argument. "Sustainable shrimps are a myth. I'm not against all prawn farming; it was part of the old ways. But then they were a luxury for local consumption. Mass production for export is disastrous. It is the scale that is the problem, and the damage it has done to the indigenous, centuries-old farming system. It was a system of survival, but it has been destroyed. We need to bring it back. The poor farmers need to take the land back and return it to paddy and mangroves." That would provide more jobs, more security—and fewer musclemen.

So I had found the alternative solution. I could see why Iqbal would have no interest in encouraging Khushi's plan. But after seeing the state of the industry that is supplying his customers, and me, it was hard not to agree with her.

Scorched Earth   6
A Sticky End with Palm Oil and Sugar

Once, our margarine was made from whale oil. This was always controversial. In the 1930s, Norwegian ports blockaded "blubber-boiling" ships owned by the food giant Unilever, demanding that they cut the cull in order to save the whale. In the 1950s, the Greek shipping mogul and socialite Aristotle Onassis was employing Hitler's old whaling captains to run the world's largest whaling fleet—and, far from being bashful about the business, he boasted to the guests on his yacht that the bar stools were covered in white skin from whale scrotums. All this so the processed blubber could be spread on our sandwiches. The slaughter was indiscriminate and all but wiped out blue whales, the largest mammal on the planet.

Thank heavens, that's over. Most whaling is now banned, and our taste in margarine has moved on too. Concern about coronaries has encouraged us to replace animal fats with vegetable oils. But now some of those vegetable oils are in trouble, blamed for rain forest destruction. More than logging, more than slash-and-burn farming, the global push to grow ever more soy in the Amazon and palm oil in Southeast Asia is wrecking the rain forests. We saved the whales, but we are losing the jungles.

I eat palm oil several times every day without knowing it. So do you. Palm oil is in an estimated one-third of all the products we pick up off the supermarket shelves. When I buy margarine or cookies, chocolate or potato chips, ice cream or commercial pastry, instant soup or noodles, or coffee creamer, I am buying palm oil. And we don't just eat it; we wash in it and adorn ourselves with it. Palm oil has replaced whale oil in my soap, and as the base lathering material in my detergents, toothpaste, and shampoo. It is in waxes and polishes, and in my daughter's lipstick. Britain alone consumes over a million tons of palm oil a year. Put personally, that works out to about 45 pounds. U.S. consumption has traditionally been lower, but is now rising fast as health concerns encourage food manufacturers to switch to palm oil in many products.

The oil palm tree—not to be confused with the coconut palm —is super-fecund. The plum-sized fruit grows in bunches of up to three thousand, weighing down the squat trees. It is a native of West Africa, and was the original source of the wealth of Lever Brothers, the British wing of the great Anglo-Dutch trading giant Unilever. Even before it was making margarine from whale oil, it was establishing huge palm-oil plantations on former forestland in West Africa to turn into soaps like Lifebuoy, one of the first global brands. In the 1960s, Nigeria was still the world's dominant supplier. But back then the world produced less than 2.2 million tons of palm oil a year. Today the figure is around 30 million.

The boom has spread palm-oil plantations across the tropics, from Mexico to New Guinea. But two rain-forest nations have oiled the wheels of this new agribusiness best. Malaysia and Indonesia between them have planted an area the size of England, and account for 80 percent of global production. The Malaysian Palm Oil Council promotes palm oil as a fruit of the forests with "the scent of violets, the taste of olive oil and a color which tinges food like saffron." But the truth is that, as in West Africa, forests in Southeast Asia are often cleared to make way for palm oil. That became dramatically clear in 1998, when rain forests burned across Borneo and Sumatra, creating smog that engulfed the region. Subsequent studies showed that three-quarters of the fires had been lit by people clearing land for palm oil.

When I nibble at a cookie or open a packet of chips or brush my teeth or whiten my coffee or tuck into an ice cream, I am helping wipe out the rain forest and reduce the most diverse ecosystem on the planet to a botanical desert.

Now a new threat looms. Not content with filling our kitchens and bathrooms, the purveyors of palm oil now want to top up our fuel tanks. Technically, palm oil is an ideal source of biofuel. European Union laws now require filling stations to mix vegetable oils with regular diesel. Domestic rapeseed oil will not meet the demand. So Malaysia and Indonesia want to fill the gap. In mid-2006, the two countries announced joint plans to set aside 40 percent of their palm-oil output for biodiesel. By that they meant they will increase production by that much. The U.S. biofuels revolution has so far largely been restricted to bioethanol production, and what biodiesel it produces is mostly made from soy. But if the market grows, Malaysia and Indonesia are determined to grab a large share of the new business.

Malaysia's leading supplier, Loders Croklaan, recently doubled its global refining capacity. And Indonesia intends to expand its plantations from 15 to more than 20 million acres. Where will the world's fourth most populous nation find all that land? You guessed. Some 5 million acres will come from converting an area of rain forest in central Borneo the size of Wales into the world's largest palm-oil plantation. Friends of the Earth concluded that the farm could "sound the death knell for the orangutan and hamper the fight against climate change, the very problem biofuels are supposed to help overcome." But green business is green business.

Right now, I can't imagine a diet that would allow me to boycott palm oil. But there are many other edible oils that could be used instead by food manufacturers without returning to slaughtering the world's whales. Palm oil is used in large part because it is cheap and readily available, while there are few effective controls on the destruction of rain forests. I would, at the very least, like to be able to choose a brand that doesn't trash the jungles. But good manufacturers and retailers who promote "sustainability" in the abstract seem remarkably reluctant to reveal where their fats come from. Campaigners who have asked have met a blank wall. Manufacturers simply say that they "buy palm oil on the world market" and so do not know where their stuff comes from. See no evil, hear no evil. They could find out if they wanted.

This ethical copout, incidentally, is provided by the same band of big commodity brokers who turn up in other agribusinesses. The palm oil that oozes through my diet comes mainly from three sources. There is Cargill, which has its own palm-oil plantations in the jungle regions of Indonesia and neighboring Papua New Guinea, as well as refineries in Malaysia, the Netherlands, Britain, Germany, and the United States. There is U.S.-based commodities giant Archer Daniels Midland (ADM), which is part owner of the Singapore-based Wilmar Group, the largest exporter of Indonesian palm oil. And finally there is ever-present Unilever—producer of major U.S. brands like Dove, I Can't Believe It's Not Butter, and Ben & Jerry's—which currently buys about 7 percent of global palm-oil production.

This is an industry that appears to be out of control, where the growing corporate claims of sustainability have yet to get much beyond greenwash. The commodity is still often produced in an ethical vacuum. And we consumers are given little choice but to be part of it. Personally, I will only believe in the greening of the palm-oil companies when they announce that they will not buy the oil from Indonesia's new superplantation. Perhaps the big service stations would like to join in too.

I don't eat a great deal of meat. My home-cooked lamb and chickens (and Christmas turkeys) come from organic British sources. Not too many worries there. But my takeout chicken tikka has a different provenance. There is a good chance my tikka comes from Sun Valley Foods, a subsidiary of the giant U.S.-based agribusiness Cargill. Just outside the small market town of Hereford in the English Midlands, Sun Valley rears about a million chickens a week, and supplies caterers across Europe—including half of all McDonald's Chicken McNuggets, though I can't say I have ever eaten one of them.

Sun Valley's "vision" is to be "the market leader in the global chicken value chain." Its parent company, Cargill, is better known for selling bulk agricultural commodities like vegetable oils and grain and cotton. So why chickens? One good reason seems to be that it can feed them on one of its other commodities—soybeans. Until early 2007, Cargill was exporting some 220,000 tons of soybeans through a terminal built for that purpose at the Brazilian port of Santarém near the mouth of the Amazon. The beans were destined for Cargill's dock in Liverpool, and from there were driven to Sun Valley Foods.

This chicken run attracted Greenpeace's attention. Wittingly or not, it said, Cargill was accelerating rain-forest destruction by pro-viding a ready market for soy grown on former Amazon rain forest. And it produced plenty of evidence of soy grown on recently lost forest around Santarém. Greenpeace mounted a sustained campaign against both McDonald's and Cargill, until the burger chain an-nounced a two-year moratorium on taking meat fed on soy from "recently deforested" areas of the Amazon, and Cargill shut down its terminal at Santarém. Sun Valley remains open for business, of course, its vision intact. Only the source of the feed has changed. It's an environmental gain, but how big is far from clear.

Unzipped 7

When the Banana Lost Its Seeds
and Other Tales from the Orchards

Pity the banana. Despite its unmistakably phallic appearance, it hasn't had sex for thousands of years. The world's most erotic fruit is a sterile, seedless mutant—and therein lies a problem. The banana is genetically old and decrepit. It has been at an evolutionary standstill ever since humans first propagated it in the jungles of Southeast Asia at the end of the last ice age. And that is why some scientists believe that the banana could be doomed. It lacks the genes to fight off the pests and diseases that are invading the banana plantations of Central America and the smallholdings of Africa and Asia.

The banana needs a pick-me-up fast. But science has so far let it down. For decades, plant breeders have all but ignored it, because developing new plant varieties without the help of sexual reproduction is expensive and time-consuming. As a result, most people in the developed world eat just one variety, the Cavendish. And the world's favorite fruit—the one I eat most regularly—could be on the cusp of extinction, says Emile Frison, an old banana hand and head of Bioversity International in Rome.

In some ways, the banana today resembles the potato before blight brought famine to Ireland a century and a half ago. But it holds a lesson for other crops too, says Emile, about how the increasing standardization of food crops is threatening their ability to adapt and survive. Popular fruits are at risk more than most. Your favorite could be on the verge of extinction.

The banana is among the world's oldest crops. The first edible banana was unzipped around ten thousand years ago in Southeast Asia. Its very survival is a testament to the wisdom and inventiveness of our Stone Age ancestors. The wild banana is a giant jungle herb with a fruit that normally contains a mass of hard seeds that make it inedible. But now and then, hunter-gatherers discovered plants that produced seedless, soft fruit. And they were very tasty. Plant scientists now know that these mutations resulted from an occasional genetic accident that prevented seeds and pollen from developing normally inside the fruit. The dark lines within the flesh of an edible banana are all that remains of the vestigial seeds. So the mutant plants were sterile, but their fruits were edible.

The early farmers cultivated these sterile freaks by replanting cuttings. And so began mankind's love affair with the banana. The first banana boats took the giant herb to Africa several thousand years ago. Anthropologists believe it became the nutritional mainstay that allowed the Bantu people to colonize most of the continent. And when Europeans first went to the Americas, the banana was among the first old-world fruits that they planted in the new world.

But on this long journey the sterile, constantly cloned banana has barely changed. Today we eat the descendants of the original cuttings taken by the Stone Age cultivators, probably from somewhere in the Malaysian jungle. Normally, cultivated plants develop genetic variety through random mutations during sexual reproduction, just as humans do. This process means that different varieties develop resistance to various pests and diseases, and adaptability to stresses like droughts. Plant breeders tap into this genetic variety all the time. But without sexual reproduction to throw the genetic dice every generation, each variety of modern banana—yellow, red, and green, from big starchy ones to small sweet ones—has come down almost unchanged from a separate sterile forest mutant. Each is a virtual clone, almost devoid of genetic diversity. And that uniformity makes the banana ripe for disease like almost no other crop on Earth.

Until the 1950s, one variety, the Gros Michel, dominated the world's commercial banana business. Found by French botanists in Asia in the 1820s, the Gros Michel was by all accounts a fine banana, richer and sweeter than today's standard Cavendish, and without the latter's bitter aftertaste when green. I don't remember, but I must have eaten it when I was young. However, the Gros Michel was vulnerable to a soil fungus that produced a wilt known as Panama disease. "Once the fungus got into the soil, there was nothing farmers could do. Even chemical spraying wouldn't get rid of it," says Rodomiro Ortiz, top banana in charge of research at the International Institute for Tropical Agriculture in Ibadan, Nigeria. So plantation owners played a running game, abandoning infested fields and moving to "clean" land—until in the 1950s they ran out of clean land and had to abandon the ill-fated Gros Michel. The king of the plantations—a fruit that ruled nations and toppled governments, that brought us the phrase "banana republic"—is now just a laboratory curiosity.

Its successor, and the reigning commercial king, is the Cavendish. This is a variety from southern China "discovered" by British colonial botanists and brought home in 1828, when it was named after the English lord who provided house room for the first samples. Being less tasty than the Gros Michel, the Cavendish languished until the latter's demise. But in the 1960s, tastiness mattered less than resistance to Panama disease. The Cavendish resisted the fungus and almost overnight replaced the Gros Michel in plantations and on supermarket shelves. If you buy a banana today, it is almost certainly a Cavendish.

But, less than half a century on, the day of reckoning may be coming for the Cavendish. The plan-B commercial banana is already being stalked by another fungal disease. Black Sigatoka has become a global epidemic since its first appearance in Fiji in 1963. Commercial growers keep it at bay by a constant chemical assault. Forty sprayings of fungicide a year is typical, making the Cavendish the most heavily sprayed food crop in the world. This is not good news for the employees of the big Latin American banana-plantation owners. In Costa Rica, the second-largest banana exporter after Ecuador and the place where my bananas usually come from, women in banana-packing plants suffer double the average rates of leukemia and birth defects. Meanwhile, a fifth of male banana workers are sterile, allegedly as a result of exposure to dibromochloropropane, which is now banned, and other fungicides that are not.

Organic farmers, who use natural pesticides, are much healthier, but they face the same problems of infestation. However the banana is farmed, black Sigatoka is getting more and more difficult to control. And now comes what could be the coup de grâce. Panama disease is making a comeback in a new form—known as tropical race 4—that attacks the Cavendish with particular virulence. So far, tropical race 4 has reached South Africa, Australia, and much of Asia. Millions of banana plants have died in southern China, the Cavendish's original home. Chemical fungicides cannot control it. So, it is only a matter of time before what they are calling the banana cancer makes it to the commercial plantations of Ecuador, Costa Rica, Honduras, and Colombia.

One footprint could do it, says Richard Markham, director of the International Network for the Improvement of Banana and Plantain. "A dirty boot with a few grams of soil from an infested site in Asia planted inadvertently in a Latin American plantation is all it would take. It's just a matter of time." And when it arrives, it will do to Cavendish what its predecessor did to Gros Michel. Game over.

With most crops, such a threat would unleash an army of breeders, scouring the world for resistant relatives whose traits they could breed into commercial varieties. Not so with the banana. Because all edible varieties of banana are sterile, introducing new genetic traits to help cope with pests and diseases is nearly impossible. Nearly, but not totally. Very rarely, a sterile banana will itself experience a genetic accident that allows an almost normal seed to develop. This gives breeders a tiny window for improvement. Honduran breeders tried to exploit this to create a disease-resistant Cavendish variety.

Every day for a year, workers laboriously hand-pollinated thirty thousand banana plants with pollen from wild fertile Asian bananas. The resulting fruit, some 440 tons, had to be peeled and sieved in search of any seeds. "I'll let you guess how many seeds they collected," says Emile. "About fifteen. And of those, only four or five germinated." Further backcrossing with wild bananas yielded a new seedless banana resistant to both black Sigatoka and Panama disease. Bingo! Well, no. Western consumers didn't like the new hybrid. Some accused it of tasting more like an apple than a banana. The only buyers today are in Cuba, where black Sigatoka wiped out normal Cavendish plantations and there is nothing else on the shelves.

Not surprisingly, the majority of plant breeders have till now turned their backs on the banana and gotten to work on easier plants. Even the commercial banana companies stay away. "We supported a breeding program for forty years, but it wasn't able to develop an alternative to Cavendish. It was very expensive and we got nothing back," says Ronald Romero, head of research at Chiquita, which, along with Fyffes and the Dole Corporation, dominates the international banana trade.

Could genetic modification come to the banana's rescue? Maybe. A global consortium of scientists is trying to produce a genetic map of wild banana varieties. If they can pinpoint the genes that help them resist diseases like black Sigatoka and tropical race 4, those genes could be spliced into edible varieties in the lab. Whether we will want to eat GM bananas is another matter, but Emile sees it as the only hope for the Cavendish. Without it, the most popular single product on the world's supermarket shelves could be heading for a sterile grave.

All over the world there are fruits, nuts, and other foodstuffs vulnerable to genetic fortune. The story is usually the same. Commercial fruit growers have concentrated on a handful of varieties, discarding the others. They have bred the chosen few to maximize yield or for some specific trait that they value most. In the process, the plant's natural ability to withstand pests and disease has been undermined. Meanwhile, the genetic stores of old varieties and wild relatives alike have often been lost. Most of the time, commercial planters spray their way out of trouble. But sometimes, as when Gros Michel stumbled, the sprays prove useless and the crop is doomed.

It could happen to some of your favorites. There are six major types of pineapple, for instance. But we eat only one, the Smooth Cayenne. By neglecting the others, and ignoring the fruit's genetic base in the wild, we risk losing the genes they contain and undermining the future of the fruit. The mango is suffering similar genetic erosion. A thousand or more varieties of sweet potatoes in New Guinea are undocumented and uncollected. In the Himalayan foothills of northern India, cultivated varieties of garlic and its wild ancestors are dying out.

The farms and hedgerows of dozens of tiny Italian islands in the Mediterranean are the last refuges for many rare and ancient plants. Watermelons are holed up in Vulcano, tomatoes in Elba, and cabbages in Linosa. But as holiday villas and desertification encroach, for how much longer will they survive?

Or take the case of the world's most widely eaten nut. The peanut began in the jungles of South America. The Portuguese took it to Africa, from where it reached North America and first gained wide popularity. Today, it is not just the world's favorite nibble, but also the most important source of vegetable protein for half a billion of the world's poorest people, mostly in Africa. But cultivated peanuts have lost much of their natural resistance to disease. In an echo of the banana story, a fungus is chasing the nut across the world, and it has few genetic defenses. The peanut's wild ancestors are believed to live only in a tiny area of remote rain forest in eastern Bolivia. Researchers believe that if they can find them, they can extract genes that can counteract the fungus. But the area has been declared out of bounds to scientists because of local unrest caused by opposition to an oil pipeline through the forest. Can the peanut survive? It would make a great movie.

A few botanical Indiana Joneses are out there trying to track down the wild ancestors of many modern crops. One of them is Emile's colleague Stefano Padulosi, the world's foremost authority on rare, unusual, and plain exotic fruits and vegetables. Without him, the chic salad vegetable called rocket would still be a forgotten weed in the ruins of his hometown, Pompeii. His main stomping ground is Central Asia, the genetic heartland of many of our most familiar crops, where he tracks down both wild ancestors and the collections of traditional varieties. Soviet scientists were masters at the business of collecting obscure varieties. But many of their collections have languished since the Russians went home after 1989. And, like your grandfather's stamp collection, the fate of the plant collections is in doubt because nobody realizes their value. The loss of these plants could prove another casualty of the fall of the Berlin Wall.

The future of the apple, for instance, may now hang in the balance. Around the world, farmers have over the centuries bred about ten thousand distinct varieties. Though only around fifty are grown commercially today, many more are kept for breeding purposes. Britain has more than two thousand apple varieties, and the U.S. government and Cornell University keep more than three thousand in research orchards. But by far the world's greatest genetic resource is in the Tien Shen mountains of Kazakhstan, where wild apple woods still grow. Ninety percent of the world's apples are believed to come from parent trees taken long ago from these woods. Many apple trees with potentially invaluable genetic traits are still in these hills. Or were when Stefano last looked. They could have been chopped down for firewood by now.

Stefano is also concerned about what has happened to the watermelons and pistachios that once grew wild across Uzbekistan, and the native walnuts of Kyrgyzstan, not to mention the equally prized forerunners of modern apricots, peaches, and almonds in their homeland of Afghanistan—a country where protecting wild genes does not have the highest priority right now.

Quixotically, perhaps, I am most interested in the fate of another native of Central Asia, the pomegranate, one of the world's juiciest fruits and prized for its exceptional nutritional qualities. Some say it fights prostate cancer. I enjoy its taste but, to be truthful, what interests me most is the prospect of one day going to find its genetic homeland in one of the world's oddest and most inaccessible countries.

Turkmenistan was, until his recent death, the fiefdom of Turkmenbashi, an eccentric leader of the former Soviet socialist republic. Once off the Moscow leash, he became an increasingly paranoid and megalomaniac leader of the independent state. Such was his omnipotence that he renamed the days of the week after members of his family and on a whim banned men from growing beards and anyone at all from sporting gold teeth. He prevented all access to the World Wide Web, shut down most of the country's universities, uprooted the state botanical gardens, and cut off funding for the country's other plant collections. Which left the pomegranate in the lurch.

People have been growing pomegranates in the remote valleys of the Kopet Dam mountains of southern Turkmenistan for six thousand years. While other countries grow pomegranates, the assemblage of ancient varieties is found only in Turkmenistan. In recent decades most of the old varieties have been lost from the country's orchards. Only around fifty are still grown. But on the edge of the mountains, starting in the 1930s, Soviet scientists assembled a unique collection of more than one thousand varieties of pomegranate trees at the Garigala experimental station. It is the holy grail of pomegranate biodiversity.

How is the collection doing? Few people really know. Most of the varieties have never even been catalogued, says Stefano. Garigala has been all but impossible to get to for some years. The last curator was Russian-born botanist Grigory Levin, who spent much of his life nurturing the collection, but eventually fled to Israel. He keeps in touch with the demoralized and frequently unpaid staff. "Many of the trees are being plowed under to make way for vegetables," he says.

The world's pomegranate collection is expiring. But the fruit could still survive. For Grigory says that the Kopet Dam mountains have one last treasure. Somewhere up there is the world's one and only wild pomegranate forest. Still flourishing, it is said. I want to walk through that forest, pick some fruit. Just for the hell of it. And now that Turkmenbashi is gone, I may get my chance.

Montezuma's Magic　　　　8
How Joseph the Cocoa Farmer
Became an Unlikely Green Warrior

Before ransacking the Aztec court of Montezuma in Mexico, the Spanish conquistador Hernando Cortez noted that the king "took no other beverage" except a drink made with a mysterious bean, the cocoa bean. He noted too that the king's subjects rarely got a taste, because a humble mug of drinking chocolate—albeit augmented with vanilla, spices, and honey, and sometimes a local hallucinogenic mushroom—was seen as the food of the gods. Chocolate in sixteenth-century Central America was a great delicacy. And it is a shame it didn't stay that way.

We probably have a Brit, Sir Hans Sloane, to blame for turning Montezuma's bean into the most popular confection on the planet. In the late seventeenth century, having seen Jamaicans preparing a chocolate drink similar to Montezuma's, the explorer came up with the idea of mixing it with milk and marketing it as "Sir Hans Sloane's Milk Chocolate," touted for "its lightness on the stomach and its great use in all consumptive cases." Sloane, who later gave his name to a famous square in London, also brought opium and cannabis and Chinese rhubarb to Britain, and created a treasure trove of foreign delights that formed the basis for the British Museum collection. But none captivated the Western palate like milk chocolate.

From Sir Hans Sloane's milk chocolate, it was a small step to Cadbury's Dairy Milk and the chocolate bars of Mr. Nestlé and Mr. Hershey. The cocoa bean from the forests of Central America became one of the world's most profitable and addictive commodities, loved by hundreds of millions of chocoholics round the world. The human race consumes more than 3 million tons of cocoa beans a year—about a pound for everyone on the planet. The business of meeting our predilection employs 14 million people, 10 million of them in Africa. But it has also done untold ecological damage, especially to the tropical rain forests. Cocoa, above all other crops, destroyed the jungles of West Africa, from Ghana to the Ivory Coast and Cameroon.

I went to meet one of the destroyers. Not one of the big cocoa growers, but a smallholder with an acre or so of cocoa trees that he planted sixty years ago outside Yaoundé, the capital of Cameroon. Joseph Essissima and millions like him have for years been branded as environmental pariahs, pioneers of deforestation. And yet something has changed in the jungle. Now ecologists are coming to Cameroon not to pillory Joseph, but to praise him. Joseph has not changed. But the ecologists' view of the world certainly has. Now they want to help him make more money out of his trees so that he can plant more. They say that planting cocoa could be the best way to save Africa's surviving forests. So how exactly did environmental bandits became environmental heroes, and is there any truth in the new story that farmers like Joseph could be the rain forest's salvation?

Cameroon harvests more than 130,000 tons of cocoa a year. That is not as much as the market leader, the Ivory Coast, but enough to give the country a place among the world's top six cocoa growers. In the Ivory Coast, most of the cocoa is grown on big plantations, but in Cameroon a typical cocoa farm is a smallholding close to the forests—in fact a smallholding that looks rather like a forest itself.

Joseph's farm is dark, dank, and full of life. There are cocoa trees, but also many others, some natural and some planted. He led me in. When he originally cleared the forest, he kept plenty of the natural trees to provide him with timber, fruit, and bark for local medicines, and to give shade to his growing cocoa trees. Most of his neighbors did the same. As we walked through his cocoa forest, Joseph pointed out each tree. Dotted around, I spotted oranges and mangoes, avocados and cherries. One I couldn't make out. "We keep this because it attracts caterpillars that we eat," he said. "They are very tasty."

With us was Jim Gockowski of the International Institute for Tropical Agriculture (IITA) in Yaoundé. "By maintaining a shaded canopy of diverse forest species, these farmers manage one of the most biologically diverse land-use systems in Africa," he said. "The cocoa plantations have more than half the species you find in a natural forest. It may not be virgin forest, but if the farmers didn't plant cocoa here, they would be doing slash-and-burn agriculture, planting maize or palm oil or turning the land over to cattle pasture."

Across southern Cameroon, large areas of former rain-forest land now lie fallow after the slash-and-burners exhausted the soils. Yet in their midst are the cocoa forests. Once dismissed as a scar on the natural landscape, they are now green oases. "The environmental benefits of a closed natural forest are now being provided by cultivated forests of cocoa and fruit trees," said Stephan Weise, the IITA station chief. He is trying to persuade landowners to convert their abandoned farmland into cocoa forest. And some large chocolate makers like Mars are funding his work.

On a small scale, Stephan is making headway. Near Joseph's cocoa forest I met Abomo, a widow who was growing cocoa trees and bananas in abandoned maize fields. Barefoot, in a vest, leggings, and a skirt, she took me around. The cocoa was for cash; the bananas were for food, but also to provide shade. She said she was one of a group of local women farmers who were taking up cocoa growing.

They are brave women. Out here on the edge of the forest, it is a long way to my local corner shop, where I may buy a chocolate bar while picking up the morning paper. And the tragedy is that at the very time when cocoa has emerged in Cameroon as an ecological crop, its profitability has slumped. The price paid for the cocoa that goes into my chocolate bar has collapsed.

I was invited to a meeting of farmers among the cocoa forests. There was a big sign chalked up in the clearing, advertising my appearance. About twenty cocoa farmers assembled in a circle. They described their business. After harvesting the beans, they broke the pods and fermented them in the fields, often under piles of banana leaves. About a week later, they carried the fermented beans from the forest to the village, where they laid them out in the sun, on tables or at the roadside. Then traveling middlemen showed up to buy the beans.

And that was where voices started to rise. Until a decade ago, the farmers explained, the middlemen had worked for a state marketing system, which bought at agreed prices and sold to international traders, under a negotiated price regime. But that system has collapsed. In its place is a free-for-all, in which the middlemen have a take-it-or-leave-it attitude to negotiating prices. With a perishable product and no storage, the farmers have no choice but to take whatever price they are offered.

The middlemen, the farmers said, sold on to even more shadowy figures in the coastal town of Douala. Frequently Lebanese, these men also acted as moneylenders to both the farmers and middlemen. The beans eventually found their way in trucks to the port of Bonaberi, where 90 percent of the harvest was bought by representatives of the trading companies Cargill, ADM, or Barry Callebaut. These three, plus Nestlé, control around half the total global market, and sell on to the chocolate manufacturers like Mars and Cadbury.

"The government used to be like a father to us," complained a tall skinny man whose sense of fair play was badly ruffled. "Now the buyers can pay what they like. We are like lambs facing a leopard. We don't know what the proper price should be. The big guys never come here, so we cannot negotiate. And the small guys...even their scales are not set properly, so we don't know for sure how much we are selling." The farmers were trying to organize themselves into a cooperative to get some bargaining power. But it was not easy.

Most agreed with the skinny man, who said, "Growing cocoa will vanish when our generation dies." Thousands of cocoa farmers were leaving their trees to decay, or clearing the forest to plant maize, groundnuts, and palm oil. There had been a big increase in forest clearing, they said. Those that stuck with the business received half of 1 percent of the price paid in the shops for a bar of chocolate.

Then they stopped to listen. I realized with alarm that they had invited me in the hope that I could offer some advice. I was from the world outside, after all. Even a local journalist had come to hear my words of wisdom. But I had to admit, I really had none. I was learning about the politics of the powerless.

While the Cameroon cocoa foresters were losing out, the only winners in Africa were larger farms that could cut costs to the bone. Most of those are in the tiny, unstable West African state of Ivory Coast, which is now the source of more than 40 percent of the world's cocoa. There, cocoa is grown in monoculture plantations with no shade and no forest cover. This produces faster instant profits, but at the expense of damaged soils, brutalized ecosystems, and the spread of crop diseases. And the crop is often harvested using contract labor brought in from the drought-hit countries of the Sahel region to the north. Frequently, I was told, labor was a payment for debt, something close to slavery. And sometimes, still, it involved children, dubbed "chocolate slaves." The traders say this practice has now been weeded out. We shall see.

Most of the beans are taken back to Europe and North America for processing. Cargill, for instance, runs huge ships carrying more than 10,000 tons of cocoa at a time into its warehouses in Amsterdam's Zaan district, which handles a quarter of all the world's cocoa. Here the beans are roasted, shelled, and ground to make a hot cocoa "liquor," which solidifies into blocks. These blocks are pressed to squeeze out the yellowy fat content, known as cocoa butter. The rest is ground again to make cocoa powder. The butter is the main ingredient in chocolate (with milk and sugar) and gives it its distinctive texture. The powder is added to chocolate bars for flavoring and is also used as powder for cocoa drinks and in food, from biscuits to ice creams and truffles to cappuccino coffee.

Not all beans are alike. And here was another bugbear. The farmers in the Cameroon forests told me their high-quality beans once attracted premium prices. Industry insiders I spoke to agreed that Cameroon cocoa, which had been introduced here by German colonists a century ago, was distinctive and once highly prized for its cocoa butter. But in recent years, the traders had taken to bulk ordering with little regard for quality. For them, a bean is a bean is a bean. So the Cameroon farmers got paid no more than the low-grade producers of the Ivory Coast.

So what should an ethical chocoholic do? Like many people, I am starting to buy more fair-trade chocolate. Like fair-trade coffee, it is providing a modest helping hand for cocoa farmers without transforming their lives or, in my view, quite providing the fairness promised on the label. (Another way to help farmers, incidentally, is to buy dark chocolate, which contains 70 percent cocoa and less sugar and milk.) But I must confess that I am still a rather unreconstructed lover of Mars bars and Cadbury's Dairy Milk. And in the end, I think that makes me part of the problem. Too much chocolate is mundane and cheap enough for us to binge on the stuff. We accept its mediocrity, and ask only for more, at cheap prices. While we do that, the big traders and manufacturers will remain in charge, and the farmers and their forests will continue to suffer.

The disconnect between my world and that of the cocoa farmers of Cameroon is just too great. One young boy, the son of one of the farmers, came up to me on my last day there and asked simply, "What does chocolate taste like?" Sadly, I didn't even have a bar to give him.

We need to rediscover chocolate as a delicacy and to find new respect for the people who grow it. There are hopeful signs. Whenever I go to the United States I can enjoy Dagoba chocolate. But my greatest delight is to go to Swanage in Dorset on the south coast of England, where I drop in at the small backstreet chocolate factory run by the Chococo company. It makes gorgeous chocolates that cost almost a pound, or around two dollars, a mouthful. Its craftsmen and women buy fair-trade chocolate from Ghana—as well as organic cream from Trevor Craig's independent dairy farm in Weymouth and butter from Bridport, both local Dorset towns. The fillings include raisins from a farmers' association in the Orange River region of South Africa, fair-trade bananas from Uganda, organic dried apricots from Turkey, honey from a Swanage beekeeper, and lemon curd from Dorset's very own Janet Pook. It's worth a pound just to read the ingredients.

Air Miles   9
Why Eating Kenyan Beans
Is Good for the Planet

Britons import a third of our food, including 95 percent of our fruit and 50 percent of our vegetables. We have relied on foreign food for centuries, of course. Oriental spices kept our food from going rancid. Tea clippers and banana boats have a long pedigree. Slave grown sugar was the foundation of the British Empire in the West Indies. And we have long feasted on Italian pasta, French wine, Dutch cheese, Spanish oranges, and German sausages. But the erosion of import controls, the reduction in subsidies for British farmers, low wages in the developing world, and cheap fuel for ships and aircraft mean it is often cheaper to buy from abroad food we would once have grown ourselves. Meanwhile, we have allowed supermarkets to hook us on year-round supplies of fresh fruit and vegetables that were once seasonal British delicacies, like strawberries during Wimbledon fortnight or new potatoes. For such reasons, our food imports are seven times greater than in 1960.

But this trend is starting to worry us. Trust in our food is being undermined by growing concern about everything from African chocolate slaves to Asian bird flu and pesticide-laden grapes from Latin America. Things may not be perfect on British farms. But how much less do we know about Bangladeshi prawn ponds and Chilean citrus orchards and Brazilian chicken factories? More than half our food imports are of produce we could grow in Britain, like California lettuces or Dutch milk or Argentinean pears. And when trucks of British poultry, beef, and milk bound for Europe drive past identical trucks bringing identical produce in the other direction, then something has gone wrong.

Sustaining each of us for a year involves the transportation of the equivalent of a 13-ton container load of food and drink for more than 60 miles. Almost one in every three trucks on British roads is carrying food. Bringing food by road to our shops emits over 10 million tons of carbon dioxide in Britain each year, and another 10 million tons abroad.

Then there are air miles. Friends of the Earth reckons that a typical stir-fry today might contain peas from Kenya, baby corn and prawns from Thailand, spinach from California, beans from Morocco, and carrots from South Africa, with combined air transportation of more than 30,000 miles. Every basket of strawberries air-freighted from California produces emissions equivalent to more than four school runs in the car. Since we import about 5 million baskets a year from the United States, that works out at 22 million school runs, or an entire year's ferrying for more than 100,000 children. Only 1.5 percent of Britain's imported fresh food arrives by plane, of which a quarter is vegetables from Africa. But that 1.5 percent produces 50 percent of all our emissions from fruit and vegetable transportation, and 11 percent of our total food transport emissions.

Sustain, a green group based in London, has been measuring how much fossil fuel energy it takes to transport our food, and comparing that with the energy we get from eating that food. It requires 66 calories of energy from aircraft fuel to bring you every calorie of food energy in a South African carrot. Chilean asparagus has a ratio of 97 to 1, and air-freighting a head of California lettuce uses 120 times more energy than you get from eating it. I can't help thinking that a sensible food industry would have got the ratio the other way round.

Organic eating, incidentally, is worse in this regard. As much as three-quarters of our organic food is imported, twice the proportion for ordinary food. Britain is a great place for growing onions, for instance. But not enough British farmers want to grow organic onions. So, more than half of those on sale in British supermarkets are grown abroad. Campaigners for organic food used to argue that energy used up in the extra food miles is generally offset by the low-energy cultivation. Energy savings from growing fruit and vegetables without pesticides and artificial fertilizer are typically around 15 percent. But one study found that, for vegetables, an extra 250 miles in a truck wiped out the difference. And, if the vegetables were being flown, then just 3.5 extra miles in the air neutralized the gains. It is perhaps not surprising that, at the time of writing, the Soil Association is considering stripping its prized organic label from air-freighted food.

But before we ban foreign food, consider this. Many of the biggest energy inputs come not from transport, but from growing and processing crops. And often British production methods are more energy-intensive. We import three-quarters of our tomatoes, mostly from Spain. In season, buying British is clearly the low-energy option. But for the rest of the year, air-freighting tomatoes from the polytunnels of southern Spain actually uses less energy than heating a British greenhouse. Likewise, imported New Zealand lamb has only a quarter the carbon footprint of British lamb, even after the meat has made its journey across the planet.

What counts is the total carbon-intensity of agribusiness. A Swedish study looked at the energy needed to bring you a McDonald's cheeseburger. It found the bun required nearly 500 calories; the 3-ounce meat burger typically required 2,000 calories (mostly from fodder production for the animal); the leaf of lettuce needed anywhere from 20 to 1,000 calories (depending on whether it was grown in a greenhouse and on whether it flew to you); the onions and cucumbers were low energy items; but the cheese racked up another 200 calories. The total for the entire burger could be as much as 5,000 calories. Substantially more than the energy to be gained from eating it.

Often, the cooking is critical. A study of a typical roast chicken dinner found that home cooking required 46 percent as much energy as breeding and feeding and getting the bird to your plate. By this reckoning, incidentally, ready-cooked chickens are the low-energy option, cutting the energy used by three-quarters.

This is an inexact science, of course. Should such calculations include the electricity used to heat or air-condition the restaurant, or even the gasoline to drive the farm laborers to work? And does it matter how the energy is produced? If a wind farm is generating the power to grow tomatoes in a greenhouse, does that make it okay?

Many people instinctively feel that air-freighting of food to Britain should be shut down. Buy local, they say. We need "food patriotism" says Conservative leader David Cameron. And several UK supermarkets are sticking labels on air-freighted produce, so that customers can choose whether to buy or not. Fair enough. I am all in favor of informed choice. But my own view is that all this math raises as many questions as it answers. And for me, just as important as any precise measure of carbon dioxide emissions is the human question. Who gains and who loses in the transactions that bring your food to your plate?

The London think tank the International Institute for Environment and Development (IIED) says that over a million farmers' livelihoods in Africa are supported by British consumers: "Nowhere are UK consumers more persistently engaged with rural Africa than through food consumption choices." Harriet Lamb of the Fairtrade Foundation asks, "What right do we have to quibble about buying coffee or green beans from poor Kenyan farmers because of the air miles involved when we have a lifestyle many times more energy-intensive than theirs?" Good question. I went to Kenya to find out.

When I was young, green beans were an English crop. They were available only in season, and often sold in the village shop from a box that had come straight from the field. I knew; sometimes I had filled that box. Growing up in Kent, the garden of England, I often spent summer holidays picking beans. Nowadays those beans are as likely to come from Kenya as Kent. Marks & Spencer and Tesco are full of shrink-wrapped, air-freighted packs of green beans. They are on supermarket shelves within forty-eight hours of being picked.

Most of the beans come from a company called, with no trace of irony, Homegrown, set up and run by a British-born Kenyan called Dicky Evans. Homegrown is a major source of winter vegetables in Britain. Dicky flies about 13 tons of beans to Britain each day, as well as zucchinis, baby corn, and sugar snap peas. Usually the vegetables travel with cut flowers from his company's big greenhouses around Lake Naivasha in Kenya's Rift Valley. But most of the beans come not from big company farms but from the fields of about a thousand smallholders.

To find them, I drove two hours east of Nairobi to the sandy hills of Machakos. Trees were everywhere as we wound up narrow roads with farms on either side, many supplying Homegrown. Eventually, we met Jacob Musyoki. Jacob's farm is two terraces dug into the bottom of a steep river valley. They cover about 2.5 acres, the size of a soccer field. Furrows distribute water from a small dam farther up the hillside.

In much of Kenya, farming is an old man's occupation. Young men find jobs in the cities. But not here. Jacob is in his late twenties and he sees his future here. "Homegrown guarantees me forty-five Kenyan shillings [about 70 U.S. cents] for a kilogram of beans. That is twice what I can usually get from local traders. And the local traders are not always here. Before I joined Homegrown, I made three thousand shillings a month; now it is twenty thousand most months. Even in bad times it is eight thousand." Around here, he told me, farmers are returning to the land, joining the waiting list to supply Homegrown.

Jacob is not rich, but he clearly reckoned he was on to a good thing. His house had a concrete floor, which is still unusual in these parts. He didn't have a car or a motorbike, but he did have a TV, which allowed him to watch English soccer. He was wearing an Arsenal cap, so I knew where his allegiances lay. He had two children in primary school, and a wife with expectations. "Ladies like to be poshed up. She wants my money to help her keep looking young," he told me with pride. He had paid a dowry for her with profits from his Homegrown sales.

As we walked his fields, Jacob told me his beans took eight weeks from planting to first pick, and harvesting ended after another four weeks. Then he started again. In a good year, his 2.5 acres could deliver 4 or 5 tons of beans, worth about $3,000. Half a century ago, this area was reckoned to be on the verge of turning to desert. So I asked how the soils were faring. He said he watered the crops three times a week and figured the soil was as good as when he started.

In return for Homegrown's good prices, he had to meet the company's requirements, which could be unpredictable, he said. Right now they were demanding 55 pounds a day from him, and that was hard to deliver. He did not know the reason, which was that it was November and there had been a cold snap in Britain. People wanted cooked beans and not salad vegetables. But it is a fact of life here that the customer is king. Output has to be geared to the UK weather, whatever the weather in Machakos. The orders coming daily from Tesco and Marks & Spencer, Homegrown's biggest customers, had to be met. Come drought or high water, Jacob had to deliver.

Jacob had fulfilled his order for the day and carried his crate from the fields to the road by his house. This is hilly terrain. The house was 200 yards up the hillside. I was exhausted by the climb, so we cooled down beneath his mango and banana trees, and checked the cows he kept to provide manure for his fields and milk for his children. He showed off the padlocked shed where he kept his pesticides, and the sealed pit for disposing of waste chemicals. "I built all this myself," he says. "Homegrown won't take us on until we have done it all."

Homegrown's smallholders are organized around some two hundred local collection depots known as sheds, where trucks arrive daily to pick up the beans and other vegetables. Jacob is part of the Umwe self-help group, who work within a mile or so of the Umwe shed. There used to be twelve farmers in the group but six left, said Jacob, because they could not meet Homegrown's standards. At the shed, Jacob stashed his beans in a store where the air is kept cool, to prevent the beans from wilting, by soaking a charcoal lining with water. It looked a little makeshift, but the technique dates from British colonial times, and every Homegrown shed has one. Later Jacob sorted and graded his beans. And at around 3 p.m., the truck arrived to collect the shed's twelve boxes.

One of the surprises at the shed was that all the farmers had mobile phones. They were part of the modern world of just-in-time retailing, and couldn't afford to be out of touch. Homegrown dictates the farmer's individual planting program, based on expected demand. But the supermarkets, and their customers, are fickle. Week by week, the planned harvest is refined. Every shed is told each night how many beans to have ready for loading the following afternoon. And as late as noon on the day of collection, the retailers may call and alter their order. On occasions, they have demanded 50 percent more beans.

Jacob's day-to-day "controller" is the shed's young technical assistant, Patrick Kinyua. He tells the farmers what to plant, when to plant it, when and how much to spray and fertilize, and when to harvest. Patrick showed me the files for each farm. I asked when Jacob had planted and fertilized and sprayed his crop. It was all there, as well as the timetable for future spraying. How much irrigation water had Jacob applied? Patrick had that, too. The records also covered pests found on the beans and any contaminants in the irrigation water. Some farm files noted boxes returned by Homegrown. One was rejected for containing beans with "curved ends"; another because they "appeared dehydrated"—the latter, said Patrick, because the charcoal cooler had not been kept wet.

The shed hygiene files contained records of everything from the thrice-daily cleaning of the toilet to the results of the farmers' regular blood and stool tests—aimed to detect diseases or any trace of pesticides in their blood. All this is demanded by British retailers before they will buy from Kenyan farmers.

I visited Umwe with John Simeoni, who was in overall charge of the Homegrown smallholders. John is a maverick. His father was Italian and his mother from Mauritius. They washed up in Kenya, where John was one of Homegrown's first independent farmers. Evans quickly spotted his flair and asked him to build up the smallholder network, which he has done almost single-handed.

Traceability is the key in modern retailing, John said. If a customer anywhere in Britain returns to the store complaining about a pack of shriveled beans, if M&S (Marks & Spencer) analysts discover a toxic pesticide, or—the real fear—if half a town goes down with diarrhea after eating Kenyan beans from a local supermarket, then, said John, "We can source any bean back to one farmer, and usually one field, within an hour. It's bureaucratic. But it's the difference between being able to sell to supermarkets or not." The system was last used, he said, when pesticides in some beans were found to have drifted from spraying on a field of tomatoes next door to a Homegrown farm.

I spent some time with Homegrown's smallholders in the hills of Machakos. I liked them. They seemed independent, cheerful, and diligent. They regarded Homegrown as good business. The people in this area are mostly Akamba, former cattle herders who took up farming when their land became overgrazed in the 1930s. Some tribal leaders had served with Commonwealth forces in India during the Second World War, and they came back with new ideas about how to protect their land. Back home, they started digging terraces on crumbling hillsides, capturing the rain that fell on their fields, and planting trees. The results were startling. Since then, the Machakos district has become a textbook example of how to regenerate land suffering from "desertification." Jacob and many other Homegrown smallholders are the heirs to what some call the "Machakos miracle."

Homegrown has also organized smallholders to grow beans north of Nairobi, on the road to Thika. Here Kikuyu farmers work the land. While the Akamba farms still seem remote and their connection to an international retailing network almost miraculous, here there is abundant evidence of globalized agribusiness. We passed a Del Monte pineapple farm where more than five thousand people harvest and can more than 300,000 tons of fruit a year, and a large tobacco-threshing plant run by British American Tobacco. Up in the hills were tea plantations, mostly supplying Unilever's ubiquitous Lipton brand, and coffee growers for Starbucks.

Here, Homegrown contracts some bigger bean farmers. Potbellied John Macharia had a 50-acre farm that was rough grazing before Homegrown came along. He had a nice house, a car, children through university, and a grown-up daughter set up with her own beautician's shop in Thika. All this, he said, was built on the profits from supplying Homegrown. He had fifty laborers to do the work. He paid them 120 shillings—almost two dollars—a day, and provided barracks lodgings that he admitted were substandard and needed upgrading. But even so, he grumbled about the labor rules imposed by Homegrown. One is that no child under eighteen should work on the farm. "How can kids learn to farm if they don't work in the fields when they are young?" he asked. "They will all go to the cities and we will lose the ability to farm."

I didn't take to him. But down the road I met Margaret—a feisty, get-up-and-go entrepreneur. While running the farm, she was supporting five children in school. She had a husband, too—a schoolteacher who "didn't bring in much money." Margaret had started just thirty months before with less than 3 acres. Now she was renting more than 7 acres, and grew more than a ton of beans a month. "Homegrown takes everything I can give," she said. "I want to rent more fields, so I can sell them more." She also grew bananas, mangoes, and oranges to feed her family and to hold water in the soils during droughts. And as a treat, she had bought herself a goat for Christmas. The technical assistant at the local Homegrown shed was in awe. "Margaret's farm is the most productive here because she is a good manager," he said. "She does everything on time. She is new but of the thirty-two farms I oversee, hers is the best."

Back out on the road, John Simeoni was growing concerned. M&S was calling in demanding more stock for the next day. M&S did this a lot, he said. "They don't hold any stock overnight. So they are constantly changing their order on the day." Evidently, the company's desire to provide its customers, like me, with the freshest produce was making life harder back up the supply chain. Knowing his regular smallholders would not be able to meet the order, John started making calls to his B list. These farmers were not yet guaranteed sales to Homegrown but were likely to get the nod soon, if they played ball.

But one of them wasn't playing ball. As we returned to Nairobi, he told John he had a better offer that day from one of the freelance buyers who travel the villages operating for other companies. He wanted John to match it. John angrily accepted. He needed the beans. But as he flapped his phone shut, there was one farmer fewer on the B list.

Back at the packing plant on the perimeter of Nairobi airport, the 180 workers on the bean line were ending their day. Mostly women, they had been at work since 7:30 in the morning, topping and tailing beans, disinfecting them in a chlorine bath, and then weighing, checking, bagging, and labeling them. Much of the work was done by hand, but some things were automated—like the machine that measured individual beans as they came down the line and picked them out in the right combinations, so that each pack weighed the same.

This wasn't work you would choose. But more than 80 percent of the workers were permanent staff, which is higher than in much of the UK food-processing industry. And wages were several times the Kenyan average. There could be a lot of overtime, which was a problem for the many women with children. And since final orders from the supermarkets didn't come till noon, when the women arrived on company buses from their homes in the morning, they didn't know when they would finish. The day I was there, it was about 5 p.m., but it could be much later. "We have rules about not requiring more than twelve hours' overtime a week, and not working after dark," said Thomas Frankum, the young Brit who ran the plant. "We sometimes breach those rules, because we have to fulfill the orders." The simple fact is that women here get home late from work, tired and running risks in the dark after they leave the buses, so that bean buyers in Britain can keep their beans a day longer in the fridge.

As the sun set over the airport, the trucks were returning from the sheds up in the hills, packed with vegetables for processing the next day. Meanwhile, the previous day's harvest was ready on pallets for the midnight flight to London. After landing, the vegetables go to a large Homegrown warehouse in Stevenage in Hertfordshire, before delivery direct to stores.

I found the Homegrown smallholder operation fascinating. Yes, it could be onerous—partly because of the demands made by UK supermarkets. But a smart group of committed young farmers equipped with mobile phones and advised by trained fieldworkers has made this method of supply a realistic alternative to large plantations. I spoke to Richard Fox, Homegrown's CEO, at his offices in a new commercial development out by Nairobi's old racecourse, a relic of the days of empire and the shenanigans of the decadent Happy Valley set. Nobody would deny that his first motive was to make a profit. But he was justified in saying that "in effect we are running a fair-trade operation. We don't pay premiums to communities, but we do guarantee over-the-odds prices to ensure our supplies. Through us, many small farmers get access to foreign markets for the first time."

Back home, I told this story to "greens," suggesting that they could buy Kenyan beans with an easy conscience. "I don't care what you say," said one. "The business is not sustainable. Just look at the air miles." Well, okay. If British customers decide they don't want to buy the green beans with the label saying they flew to the supermarket, then Homegrown's business is doomed.

On the face of it, the statistics don't look good. Emissions from air-freighting beans are two hundred times greater than if they had come by ship. But the food-miles issue isn't that straightforward. In summer there are green beans available grown outdoors in Britain, and eating them is the low-energy option. But the energy needed to air-freight vegetables from Kenya to Britain in winter, when British demand is highest, is actually only about 15 percent more than the energy needed to heat a greenhouse to grow those vegetables here.

In any event, there is an issue of equity here. Can it really be right to try to make a tiny reduction in our own emissions by depriving Kenyan farmers of their livelihoods? Think of it this way. The average Kenyan's carbon emissions are one-thirtieth of those for the average Briton. For the sake of argument, let's say Jacob should be personally responsible for the entire carbon footprint of his business, right to the supermarket shelf in Britain. How do things look?

Every kilogram (about 2 pounds) of Jacob's green beans flown to Britain consumes half a gallon of aviation fuel, which releases 4.25 kilograms of CO_2. So the carbon footprint of Jacob's typical annual production, taking off a bit for topping and tailing and wastage at the Kenyan end, is about 19 tons of CO_2. That is, admittedly, approaching twice the average emissions for a typical Briton. But Jacob's farm supports a family of four. So divide by four and the per capita carbon footprint of his business comes out at about half that of average per capita emissions in Britain. Surely he has some rights here? Do you still want to boycott his beans?

Many of the people who recoil at the idea of vegetable air miles, and would not be seen dead buying a Kenyan green bean in their supermarket, also want to help poor countries like Kenya through "trade not aid." There is a contradiction here. And this is not a trivial issue. Green beans are a major Kenyan export, and 70 percent of those exports come to Britain.

My skeptical green friends point to other "sustainability" issues. One is water. Green beans imported to Britain soak up, by one estimate, more than 160,000 acre-feet of water a year. That is enough, it is said, to provide 10 million Kenyans with daily household water, in a country where millions do not have running water in their homes. Maybe so. But the Kenyan drinking water problem is largely about pollution and poor infrastructure, not an absolute shortage of water. Shutting down the farms of smallholders growing beans for Britain would not put water in taps.

Still tempted to reduce your emissions by cutting out Kenyan beans? Take the bus to the supermarket, buy a bit less processed food, do something that will hurt you, yes you, not Jacob. As a result of what I saw on my trip, I decided to start eating more Kenyan beans. And back home, I did just that, buying them at my local M&S. The beans were certainly fresh. They lasted for a week in my fridge. Thomas had promised that if I sent him the tracer code stamped on a pack, he could download the information available on who had grown it. So I e-mailed him the code: 061104005. Just that.

Good as his word, a couple of days later Thomas replied. My beans had been grown by Mary Lenkiyieu, who has a tiny farm—more a vegetable garden—covering only about an acre. It is in Loitokitok, right on the border with Tanzania. My beans had been sown on September 28, 2006, eleven weeks before I bought them. They were harvested on November 28, and packed and put in the hold of one of Dicky Evans's aircraft on November 29.

The records revealed that, along with the beans, Mary grows maize and tomatoes and beans for her family. She was investing her income from Homegrown to buy egg-laying chickens and a cow to provide milk for her young family, as well as school uniforms and books.

When I started on this journey, I thought that at the end I might concoct an elaborate allegory. Jacob and the Beanstalk, maybe. It seemed quite neat. A generation ago, Jacob Musyoki and his fellow Akamba in the Machakos hills swapped their mothers' cows for beans. The beans grew. Jacob gained some modest riches. Metaphorically at least, he climbed the beanstalk and found himself in a strange land of just-in-time sales to European supermarkets. But then the allegory fell apart. For now, there just does not seem to be an ogre to chase him out of the magic kingdom. Unless, of course, it turns out to be the air-miles crusaders.

Part Three

My Clothes

Drought and Dirty Secrets in the World of King Cotton

10

Geoff Hewitt has a big new house and a four-car garage. He has 5,000 acres of cotton farm and enough trees down by the creek for the roosting kookaburras to wake him in the morning. He lives at Macalister, near Dalby on the Dalby Downs, a four-hour drive inland from Brisbane, the capital of Queensland. People like him made modern Australia: restless, purposeful, a touch idealistic, ever ready to open his door to a visitor.

Out here there are cattle, grain, coal mines and, usually, cotton. But outside on the veranda, listening to the kookaburras laugh, Geoff was explaining to me why, months into the growing season, not a single square yard of his sprawling cotton farm was planted. Why his high-tech, GPS-navigated tractors, able to steer to an accuracy of less than an inch, were idle. And why his plan to sell branded Aussie cotton shirts in British stores was on hold. No rain. It was as simple as that. No rain, so no water in his reservoirs. So no point in planting cotton. "We've had three and a half inches of rain so far this year. Usually we get twenty-five inches in a year, and it's September."

For in 2006 the driest continent was in the midst of its worst drought in a thousand years. Nobody knew if what was happening was natural variability or a product of manmade global warming. By 2007, climatologists were saying whole towns might have to be evacuated. And already a whole industry—the cotton industry, which in a normal year takes a tenth of the country's water—was on its knees. There was no water to take.

Geoff couldn't have been closer to the heart of the crisis. As we looked out across his parched fields, he pointed toward a dried-up creek. "That is the start of the Condamine River," he said. "It is one of the sources of the Murray-Darling, the largest river system in Australia." The Murray-Darling drains an area twice the size of France, flows all the way to Adelaide, and irrigates 60 percent of Australia's agricultural output. In the good times, four-fifths of the water on Geoff's farm comes from the Condamine. But he is allowed to pump only when there is enough water in the river. And in September 2006, there was no water. So his reservoirs, which can hold more than 4,800 acre-feet, were all but empty.

Geoff hoped for the best. "This is a land of extremes. We are ready to ride out the tough times and seize opportunities when they present," he said. But he was fearing the worst. "We've not had a flood here in ten years. I don't know when we'll see another." And for now he was making money by hiring trucks to take coal from the mine over the hill to the railhead.

Aussies are smart and resourceful. But farmers had been selling up on Dalby Downs, said Dan Hickey of Cotton Australia, a service agency for the cotton farmers. Dan spent his days driving across these hills and he had fewer farmers to visit each time. A decade ago, there were 600, he said. Now there were only 350. Some had had only a third of their normal allocation of irrigation water in recent years. Some were getting none.

And yet the farms I saw still seemed oddly wasteful of water. Geoff had some of the most sophisticated, customized, computerized methods available for delivering just the right amount of nutrients to soils. But I was surprised to see that he didn't use a drip irrigation system. He said the cost of installing drip irrigation would be too great and the benefits too small. "Even with drip, we would still have had no water for the past couple of years, and the extra cost burden would have put us out of business," he said.

But the biggest losses were from his reservoirs. Geoff captured all the rain that fell on his land. Yet, he admitted, "we lose about 5 feet of water to evaporation from the surface of our reservoirs in a year." And, despite having huge amounts of detail on other technical aspects of his farm, he couldn't tell me how much water he lost from his reservoirs. His largest reservoir, covering some 75 acres, had a bank only 4 yards high. I calculated that, depending on how much rainfall he had, he must lose between a tenth and a third of his water to evaporation. He agreed that was "probably pretty close to the mark." The Australian cotton industry is always a whipping boy when the rivers run low. The farmers are aggrieved. But I didn't find it surprising. Just 1,500 cotton farms use more than the country's 7 million householders. And they only seem to care when the water runs out.

Geoff's farm looked pretty big to me. But it was tiny compared to what, so far as I can establish, is the world's largest cotton farm, a couple of hours down the road near Dirrabandi on the border with New South Wales. You can spend a day just driving round the edge of Cubbie station. The entire farm, which until twenty-five years ago was a cattle ranch, stretches 25 miles in one direction and 12 miles in the other. Nearly half is now irrigated. Its reservoirs, which like Geoff's take water from the Condamine, can store more water than Sydney Harbor. The thinking is not that the reservoirs will fill every year, but that when the big flood comes (if it comes), the farm can take all the water available. Then they hope they can use most of it before it evaporates.

Station manager John Grabbe has a simple philosophy. "I firmly believe we have a responsibility to absolutely maximize production from the water we are entitled to divert." He is as good as his word. After a program to reduce evaporation by deepening his reservoirs, he followed up not by giving water back to the river, but by increasing the acreage under irrigation by 50 percent. Grabbe complains about being singled out. He takes only 15 percent of the flow of the Condamine, he protests, and less than 1 percent of the entire flow of the River Murray system. But that is still a lot for one cotton farm.

I had planned the trip to Dalby Downs expecting to see the cotton that made my shirts growing in the field. That plan was derailed by the drought. But everybody told me that normally there would have been cotton that would end up in clothing sold in my local M&S or other mainstream clothing stores across Europe and North

America. Geoff said, "We sell cotton from this farm to Paul Rein-hart, who I know sells our quality cotton to M&S. Every bale is tagged from this farm, so we should be able to trace it." Cotton Aus-tralia had gone further: "Together with Marks & Spencer, we have traced out the supply chain from gin-yards in Australia to spinners, knitters and weavers in Indonesia, Thailand and Korea, garment manufacturers in China, Vietnam and Bangladesh, and from there to European and North American stores."

But when I inquired further, neither M&S's chief cotton buyer, Graham Burden, nor the Australians could be sure of the precise journey. It became a deepening mystery. For almost a year, both sides promised to try to firm up the link. I had an e-mail from Burden say-ing that he had tracked some products back to Indonesia, a major outlet for Australian cotton. He was "drilling down to the next layer" to make the final connection. But it never came in time for this book. Allan Williams of the Cooperative Research Centre for Cot-ton, which was doing the research for Cotton Australia, was work-ing from the other end. But he too hit a brick wall. It seemed that no grower knew for sure where his cotton ended up, and no retailer knew where his cotton originated.

As in other major commodity industries, like coffee and cocoa, international trade in raw cotton is concentrated in the hands of a few companies. They include Dreyfus, Cargill, and Dunavant (the largest) in the United States, Reinhart in Switzerland, Plexus in Liverpool, and the newcomer Chinatext in China. The problem, as Geoff put it, is that these traders have no interest in putting the grow-ers in touch with the markets. "They might cut out the middleman if they knew how the supply chain worked."

There is a real problem here. In some industries a transparent supply chain is emerging. But not in cotton. I felt I had a right to know where my cotton came from. And Australian farmers were similarly aggrieved that nobody knew about their quality cotton. They have a program called Best Management Practice, or BMP. Begun in the 1990s amid scandals over the misuse of pesticides on Aussie cotton farms, BMP has helped clean up toxins in the indus-try. It is, said Geoff, one of its most ardent advocates, a voluntary toolkit for farmers, covering soil and water management, chemicals spraying, and fertilizer application. He showed me the paperwork involved. Now the farmers wanted to sell BMP cotton as a brand at premium prices. Not organic, not fair trade, not GM-free—but top quality. "We want a label on every shirt," he said. It could be done, if the traders cooperated.

I had my doubts about the "sustainability" of Aussie cotton, given its clearly unsustainable water use. But anything that starts a debate on those issues sounded like a good thing. In any event, a few more years of drought and Geoff would have to become a full-time coal hauler, sell his nice house—or make better use of his water.

But I was still waylaid on my journey: unable to establish a link between my wardrobe and the cotton fields of, well, anywhere. I needed to know more about the industry. World cotton production is more than 25 million tons a year (the equivalent of fifteen T-shirts for everyone on the planet). *Homo sapiens* collectively spend more than a trillion dollars a year buying clothes, with cotton the main ingredient. This would keep the GDP of Mali ticking over for almost two hundred years. Every year Americans spend about $800 each on clothes.

Most of the cotton for all this comes from seven countries: from the large mechanized farms of Australia, the United States, and Brazil; the millions of small farmers who dominate in China, India, and Pakistan; and the huge state monopoly of Uzbekistan in Central Asia. Despite being major producers, China, India, and Pakistan export little raw cotton. This is because they spin most of what they produce and turn it into clothes within their national borders. As a result, cotton exports are dominated by just three countries: the United States, Uzbekistan, and Australia. It is these countries that supply the raw materials for the other big players—the producers of low-wage "sweatshop" garments—who take advantage of the fact that there are no automated processes that have yet been found to replace the work of human hands and sewing machines.

The cotton industry has always had a bad reputation for abuse of human rights. From the days when cotton was one of the prime products of the British Empire, and of course in antebellum America, it has been associated with slavery. And the stain lingers in places like Uzbekistan, where state compulsion to go cotton picking is routine, and in Pakistan and India, where child labor remains common. But today there is equal concern about the environmental impact of producing this apparently natural product, which turns out to be huge.

One issue is pesticides. While occupying just 2.5 percent of the world's croplands, cotton uses a tenth of all the world's chemical fertilizers and a staggering quarter of all the insecticides, mostly to fight off whitefly and bollworm. In India, half of all the country's pesticides are sprayed on its cotton fields. The health effects of this are poorly quantified but undoubtedly substantial, with thousands of field workers thought to be dying of pesticides poisoning in India alone each year. When a chemical factory burped a toxic cloud over the Indian city of Bhopal in 1984, with many thousands of deaths, it was making pesticides for cotton farmers.

Then there is the crop's huge demand for water. Cotton grows best in hot, sunny regions—typically deserts. Three-quarters of all cotton production around the world requires artificial irrigation—more than for any other major crop. From the Yellow River Valley of China to southern Pakistan, from the headwaters of the Murray-Darling to the dried-up Aral Sea, cotton empties rivers and lowers water tables on a scale achieved by no other crop.

How is this reflected in a single T-shirt? Billions of T-shirts are bought and sold every year. Britain alone imports almost half a billion of them each year (or eight per person) and America imports more than a billion, on top of domestic manufacture. They often cost little more than the price of a beer. Yet the average cotton T-shirt, which weighs about half a pound, requires about 3 ounces of fertilizer, a tenth of an ounce of active ingredient in pesticides, and between 500 and 1,800 gallons of water, or upwards of thirty bathtubs full. For one T-shirt. The market ignores this. Far from reflecting these environmental concerns in high prices, world cotton prices have been on the slide for more than a decade.

Two factors have influenced this. First, the spread of Monsanto's GM cotton, engineered to resist the bollworm. This cuts pesticide use. But, in the past five years, the increasing popularity of GM cotton among farmers has raised yields by 20 to 30 percent, flooding the market at a time when demand for cotton has been static because of growing consumption of manmade fabrics like polyester.

The world cotton industry is also seriously distorted by state subsidies for cotton farmers. Subsidies inflate production and so depress market prices. By most measures, Australia is the only country that does not supplement the income of its farmers. Elsewhere, as much as a fifth of all cotton growers' earnings comes from governments. The biggest subsidies come from the U.S. government. The largest recipient has been Tyler Farms, which covers 40,000 acres of the Mississippi Delta in Arkansas. Its owners pocketed almost $37 million in the decade to 2004. As an Oxfam study of impoverished West African cotton growers pointed out, "This one farm receives subsidies equivalent to the average income of 25,000 people in Mali."

So far I haven't mentioned cotton's carbon footprint. This turns out to be large, but surprisingly, much of it arises not from making your T-shirt, or even transporting it around the world, but from using it—more particularly from washing and drying it. Growing the cotton for that shirt generates about 2 pounds of carbon dioxide, mostly in making agrochemicals and pumping water. Turning that cotton into a shirt accounts for about 3 more pounds at spinning and knitting and garment factories. Transporting the cotton between those factories and ultimately to your home takes perhaps a further pound, though quite a lot hangs on whether you drive to the shop especially to buy it.

After that, if you are a fairly typical user, you will machine wash that T-shirt twenty-five times before throwing it away. If you wash at 140 degrees, then tumble dry and iron the shirt, it will emit about 9 pounds of CO_2. That is more than the energy used in producing and delivering it to you, and gives your T-shirt a lifetime footprint of 15 pounds of CO_2 or twenty-eight times its own weight.

We can take this further. Some analysts have put the "social cost" of CO_2—that is the amount of damage it does in changing the climate—at $250 per ton of CO_2. On that basis, a typical T-shirt causes about $1.80 worth of damage to the climate during its lifetime.

But remember that even after making your purchase, you are in charge of the larger part of that carbon footprint. This study was funded a few years ago by M&S, and it does seem designed to point the finger at consumers. Who washes at 140 degrees? Isn't 120 degrees more usual? Do we all routinely use a tumble dryer? And who on earth irons their shirts every time they wash them these days? You can cut that footprint substantially. M&S, which is now cuddling up to its customers rather than accusing them, has a full-page advertisement in my paper as I write, pointing out that by washing at 85 degrees rather than 140 degrees, we can save 40 percent of the energy bill. Hopefully, it will also soon add that hanging your washing on the line to dry will save even more.

M&S has also looked at the energy footprint of a viscose blouse. Viscose is a fiber made from wood. It takes twice as much energy to make viscose as to grow cotton, but less to turn it into a garment. So far they are even. But because you can wash viscose at lower temperatures and it drip-dries in a jiffy, the carbon footprint of the viscose blouse from day-to-day use can be as little as a tenth that of a cotton T-shirt. It might be going a bit far to suggest, as one headline writer did, that viscose can save the world. But it could help. Personally I hate manmade fibers. But I'll cool-wash cotton, I don't own a tumble dryer, and I never iron shirts, even cotton ones. So I hope I am not too bad.

Behind the Label
Bless My Cotton Socks

11

Searching for some new socks in Marks & Spencer one day, I noticed, buried way down low, a small collection of fair-trade cotton socks. They cost twice as much as the other pairs, but I bought them. The label made me feel that the other pairs were unfairly traded. In fact, I felt so good about the purchase that I went back and picked up a $16 fair-trade T-shirt. At the time of writing there are no fair-trade socks or other clothing certified for sale in the United States. So Americans can't yet join me in feeling good here, though I am sure things will change soon. But, apart from the feel-good factor, what extra was I getting for my money? More to the point, who was getting my extra money? Armed with the product codes from several recent purchases, plus M&S's store slogan promising to go "behind the label" to tell customers where their clothes come from, I talked to M&S's Graham Burden, at the company's west London headquarters.

M&S turned out to be a bit hot and cold on going "behind the label." They wanted to do it when they had a good story to tell, but not when they didn't. Graham was not keen for me to pursue the origins of a pair of $18 jeans made in Bangladesh. But he was keen to make the introductions for me to follow the trail of my fair-trade socks and T-shirt. And it is a good story.

It was still dark when I got off the sleeper train from Mumbai. Surendranagar station was in the middle of nowhere in the near-desert Indian state of Gujarat. But I stepped into a predawn hubbub. It was the end of the Eid festival, and the platform was packed with people returning from the city. I eventually found my host, Dilip Chhatrola. We took an early truck-stop breakfast with drivers wrapped in blankets against the chill, and headed north for the small agricultural town of Rapar. The highway was full of Tata trucks bringing raw materials from the huge seaport of Kundla to factories making steel girders and ceramics and much else. But as we approached Rapar, the sun's first rays caught the white of cotton in fields ready for harvest. In the town, the morning rush hour mixed bullock carts with motorbikes. A camel was pulling a cement mixer. This was more like India.

I had come to see organic cotton being grown under fair-trade rules. My contacts this cold desert morning were from Agrocel, the company training the farmers in organic ways and buying their cotton to make into my socks and T-shirt. Curiously, Agrocel's small depot in Rapar was stacked high with agricultural chemicals. They included organophosphates like methyl parathion and dichlorvos —one with the alarming brand name of Doom. These are lethal products, especially in villages where protective clothing is virtually unheard of, where farmers have nowhere secure to store their chemicals, and where many cannot read the instructions on the labels. This was the enemy, surely?

Agrocel is an offshoot of a larger Indian company best known for its pesticides. But it has been changing tack. The Rapar store manager said, "We sell pesticides, but we explain to the farmers at the same time why they should switch." It sounded like greenwash, but I was wrong. In this corner of Gujarat, Agrocel has established one of the world's largest farmers' associations for producing organic cotton, and recently won certification from the Fairtrade Foundation. In the villages around Rapar, 560 smallholders were growing organic cotton and receiving a premium price from Agrocel in return.

We set off to meet them. Each described how, after attending Agrocel classes, they had given up on that company's products and begun composting farm waste, recycling cattle dung onto the fields, and using natural plants to ward off pests. They ground up seeds from the local neem tree to make an oil that killed aphids and the dreaded bollworm, for instance. "Before, I sprayed endosulfan every week," said Amarsi Chhandha in Bhimasar village. "It cost me a lot. But neem oil is cheaper and safer and works well." Most farmers also planted crops that attracted pests away from the cotton. Marigolds were popular, and not just as a pesticide. Everywhere I went, farmers' wives garlanded me with elaborate wreaths of the bright yellow flowers.

This reliance on natural pesticides was very impressive. Whenever I had met cotton farmers before, they had told me their crops failed without a constant chemical dousing. But here they were adamant that organic methods worked. Ranabhai Dungarbhai, a numerate and articulate farmer, said, "Before Agrocel came here, I sold to local traders who paid less and often paid late. I used to get about twenty rupees for a kilo of seed cotton straight from the field. Now it is about twenty-four rupees, and the price is guaranteed." His yields had initially fallen when he switched to organic, but they recovered and were now down just 7 percent. Without pesticides to buy, his farm costs were down 20 percent. So he was better off. He planned to spend his profits on a new house for his three young children.

The big problem was not pests but water. Gujarat is dry, and cotton is a thirsty crop. One farmer in Padampar told me that his father could bring water to the surface in a bucket, but now he ran electric pumps around the clock to haul it onto his fields from 100 yards down. "We have to dig deeper every year," he said. Organic farming was helping. "The soils are better and hold more water," said Ranabhai. He used to irrigate once a week, but now turns on the pumps only every ten or twelve days. But the water tables still kept falling. One village was taking steps to increase the water supply. Touring the Padampar village, I saw ponds and earthworks dug by hired earth movers to capture the monsoon rains and pour it down wells to recharge the underground reserves.

Who was paying for this muck shifting? Agrocel. Besides the premium paid to farmers, it also pays an additional 13 percent to village associations for community projects. And the people of Padampar had invested most of the $40,000 they received the previous year in catching the rain. They could not have done this without Agrocel, they said. The community premium—equivalent to $20 per head of its population—was more money than the village got from the government.

The premium was being spread widely, however. Virtually everyone in the villages benefited. At Padampar school, teacher Babubhai Klor came in on a public holiday to show me three shiny new taps. For the first time his 360 pupils had drinking water in the school. Then he showed off the school's small herb garden. And we went into the classroom, where he opened a cupboard to reveal two small microscopes, a tray of test tubes, a torch, a pinhole camera, some tiny solar cells, and a miniature windmill—all paid for with the communal premium. I began to feel good about my cotton socks.

Agrocel's main office is at Mandvi, a four hours' drive from Rapar. There, I asked Hasmukh Patel, the charismatic head of the fair-trade project, why an agrochemical company was sponsoring organic farms? He said the parent company, Excel Crop Care, was one of the philanthropic enterprises set up by self-styled Indian visionaries after independence from Britain in 1947. Its founder, G. C. Shroff, had dreamed of uplifting peasant farmers by selling them pesticides. Back then, access to farm chemicals was seen as the route to prosperity for the rural poor. This was before Rachel Carson's book *Silent Spring*, published in 1962, sounded an environmental warning about pesticides. Nobody spoke then about a "pesticides treadmill," and the health hazards were largely unknown.

But Shroff later saw the light. So he has sent out his top lieutenants to help farmers grow organic cotton and to find markets for their produce. Bishopston Trading Company, a cooperative clothes shop in the west-of-England city of Bristol, started buying Agrocel cotton to be spun and woven on hand looms in southern India. Agrocel also got help from the Shell Foundation, a charity set up by the petrochemical company to help poor farmers with the problems of globalization. This was fascinating. Two companies whose prime business is selling agrochemicals had gotten together to found one of the world's most successful organic enterprises.

After going organic, Hasmukh spotted the growing market for fair-trade products. And now business is booming. Demand far exceeds supply. The week before I flew to India, Marks & Spencer had announced that it wanted to buy one-third of all the world's fair-trade cotton. A couple of weeks before I showed up, their local buyers had been in the Agrocel office asking to purchase more than 7,000 tons from Agrocel's 2007 harvest.

That caused a big laugh. Agrocel is growing by 40 percent a year, but the order was still four times higher than its expected production for the year. "M&S pushed and pushed," Hasmukh said. "We told them our first priority was our small, regular customers. Whole villages in India depend on our cotton to make clothes for people like Bishopston. M&S may be big, but they are newcomers." In the end, he promised to sell them 330 tons. "One day we may be able to meet M&S's demands," he said. "But not yet."

Many fair-trade cotton suppliers are facing similar dilemmas. They crave the big contracts, but fear the major brands could disappear as fast as they have arrived. And even if they stay, they may end up subverting the fair-trade formula. Hasmukh is optimistic, however. He believes that the communal premiums could be the key to long-term success. "They have really encouraged farmers to share expertise, through the associations set up to manage the money. We also find that women in the villages are taking a big part." He saw two major threats. One is GM cotton, which is spreading fast through India. The other was the shortage of water, which a surge in demand for even organic, fair-trade cotton could make worse. In that sense, the fair-trade cotton boom may contain the seeds of its own destruction.

Maybe in the long run, it is too dry in Gujarat for cotton farming of any sort. But at least the farmers seemed to be making informed choices about their water.

India is the world's third-largest cotton producer, with more land under cotton cultivation than any other country. Its big industrial combines invest in high-tech spinning, knitting, and weaving plants. Welspun is the world's biggest producer of fabric for babies' diapers. Arvind in Ahmedabad is one of the world's biggest manufacturers of denim, churning out a staggering thirty thousand pairs of jeans a day for brands like Levi's, Lee, Diesel, and Wrangler. But I was following my personal footprint, the trail of the cotton that went into my fair-trade socks and my fair-trade T-shirt. And that took me east to the state of Madhya Pradesh, in the heart of India's cotton belt.

Here, on the banks of the River Narmada, among traditional handloom weavers and thousands of sari stores, I visited Maral Overseas, a large integrated cotton operation covering hundreds of acres. Maral is a global company, growing its own cotton, importing more from across the globe, and employing thousands of workers to make yarn and clothing. It turned Agrocel cotton into yarn for my socks, and produced my fair-trade T-shirt from start to finish.

Maral, like Agrocel, has a tradition of Hindu philanthropy. Its campus has free food and clinics open to the entire community. The on-site farm grows vegetables and fruit for the factory kitchens, and has a herd of cows living in a cowshed as well air-conditioned as its guest rooms. The cows supply the canteens with milk, and their dung fertilizes the vegetable gardens. Every cow has a name: Yamuna, Kamini, etc. The bull doesn't.

Maral funds a local school that is so good that children come from 25 miles away to attend. I was introduced as a guest of honor at one morning assembly by the principal, Mr. Panda—a small, bustling, improbable man with a sharp business suit and a punkish quiff of hair. The uniformed students stood smartly to attention. It all felt like a scene from a Gilbert and Sullivan opera.

The company also has 1,200 farmers of organic cotton under contract and produces good yields. This, incidentally, gave the lie to claims I heard elsewhere that Agrocel's organic cotton farming was only possible because of some peculiarity of the local cotton. The project's mastermind, Surendra Kumar Tiwari, took me around the farms. This was no monoculture. Interspersed with the cotton, I found parsley and aniseed, coriander and eggplant, chilies and ginger, papaya and tomatoes, runner beans and gilki (which is like a zucchini), red onions and chickpeas and turmeric. There were Mahua trees, whose flowers produce a traditional alcoholic drink. The farmers had planted lemons and potatoes around the fields to attract birds that eat the pests.

As we sat beneath his mango tree, farmer Sitaran Giadi twiddled his fine mustache and told me profits were good. His children go to university. "I left education when I was thirteen," he said wistfully.

Other farmers had sent their children off to engineering college, or bought more cattle or dug more wells or arranged better marriages for their children. "Living standards have improved since we went organic," everyone agreed at an outdoor meeting. Most now had TVs in their homes and about a quarter had motorbikes, though mobile phones were still rare.

One farmer's wife took me into her bedroom to show me a small wooden chest where she kept gold jewelry. It was clearly like a bank to her. It reminded me that the people of rural India have, collectively, the largest stash of gold in the world. Cotton, the "white gold," was paying to buy real gold.

Maral is a huge operation. Less than 1 percent of its cotton comes from these farms. And as M&S's biggest supplier of organic and fair-trade clothes, it shares in the problem of where to find the certified cotton to make those clothes. Agrocel cannot meet the demand, so they are going farther afield. The Maral warehouse turned out to be full of bales from Cameroon, which has established a beachhead for fair trade. Maral had recently imported more than 1,600 tons of fair-trade cotton from West Africa, enough to meet M&S's needs for six months.

Maral produces about 40 tons of yarn a day, a big part of which is bought by M&S. Most of the operation is automated. But even so, more than six hundred people work in the four spinning sheds, cleaning the cotton, combing and compressing and then spinning the fiber, and loading it onto giant bobbins, each with more than 60 miles of yarn. Maral's forty knitting machines turn 10 tons of this yarn into fabric each day. The fabric is washed and dyed and cut into shapes on huge cutting machines. Then, for the first time, I heard a hubbub created not by machines but by human voices. Some nine hundred people on two shifts, armed with sewing machines and shears, turned those oddly shaped fragments of fabric into clothes, like my fair-trade T-shirt.

There were some pretty unpleasant jobs here. I watched women making buttonholes all day. Then I found a small side room where a dozen young men were sitting cross-legged on the floor separating collars that had been made in bulk by machines. They pulled a thread and tore a collar away, time after time after time. Afterwards, there was ironing and folding and tagging and packing.

Maral sells branded goods to U.S. companies like Nike, Columbia Sportswear, Timberland, and Marlboro Classic, as well as Canadian firms like Cotton Ginny and Hudson Bay, and to M&S in Britain. My fair-trade T-shirt was made here in its entirety, put on a hangar in the dedicated M&S storage area, and loaded into a container for the journey to Mumbai port and London. But it turns out that Maral has only a bit part in the saga of my other fair-trade item, my socks. There is a long way still to go to my sock drawer—a journey that made me ponder whether Maral's role was altogether sensible. This fair trade ultimately seemed pretty mad.

My socks, produced several months before my visit, had been made with Agrocel cotton. But by the time of my visit, Agrocel was being overwhelmed by orders. Because Agrocel could not supply enough cotton, most of the fair-trade socks by then in stores contained cotton grown in distant Cameroon. That cotton was trucked to Cameroon's coastal port of Bonaberi (the same port that exports cocoa for my Mars bar). Then it was shipped around the Cape of Good Hope to Mumbai, for a sixteen-hour drive to Maral, along 300 miles of road that are among the busiest and scariest in India.

At Maral, the African cotton was spun into yarn and then put back in a truck and driven all the way back to Mumbai, where an Israeli company called Delta Galil took charge, as the yarn went on another long sea journey, this time through the Suez Canal to Turkey. Near Corlu, west of Istanbul, a firm called Aloha dyed the yarn. After that, Delta took it back to Istanbul and knitted it into socks, before trucking those socks across Europe to Delta's warehouse in Northampton. That was a journey of maybe 7,500 miles. All for a pair of socks. As this book goes to press, every pair of M&S fair-trade socks makes that same journey.

What lies behind these convoluted journeys? Price, of course. It makes economic sense to ship these products back and forth across the planet in order to find the cheapest place for each stage of manufacture. Maral is competitive in some areas, but not in others. You see, hard though it may be for us in the West to believe, Indian labor is comparatively expensive in the cut-throat world of global textiles, and it cannot compete for many labor-intensive activities.

Maral is a technically very efficient plant. It can hold its own in the high-tech carding, spinning, and knitting operations. It is a cheap place to make yarn. But other stages in the process use more labor, and Maral pays its unskilled workers 120 rupees for an eight-hour day, and its skilled workers may get double that. That works out at between 35 and 70 cents an hour. And at those wages, it cannot compete these days with sweatshops in some other countries, like China and Bangladesh.

Maral can hang on to some fair-trade and organic goods, like my T-shirt, because customers will pay more and the extra wage costs can be absorbed. But for the rest, for cheap, throwaway fashion, it is in trouble. Its management told me that three-quarters of its yarn, whose production is automated, is shipped out to sweatshops for the labor-intensive activities involved in turning that yarn into garments. They just could no longer compete for that business.

Unwilling to pay Maral's wage rates, the big brands in Europe and North America are all switching much of their business to the very cheapest factories paying the very lowest wages. "H&M buys mostly from Bangladesh now, and Marks & Spencer is increasingly doing the same," Maral's managers told me, along with U.S. names like the Gap, Wal-Mart, Sears, and JCPenney.

Even M&S? That company used to have a reputation for buying where quality was best. But twice in recent years it has come close to collapse because of uncompetitive pricing. So it has transformed itself into a prime player in what economists call the "race to the bottom." When price is all and the customer wants a $6 T-shirt, then, if the cutters and sewers and ironers and packers come cheaper in Dhaka, that is where they will be employed. So, fascinated as I was by my fair-trade organic detour, it was Dhaka where I went next.

Trouser Truths

The Unscrupulous World of Sweatshops

12

Five women lived in the room altogether. Three of the women, Aisha, Akhi, and Miriam, lined up on one bed. I sat on the other. There was just space between the beds to dangle our legs. The roof and walls were corrugated iron. There was a black-and-white TV flickering in the corner and a fan overhead to make the hot evening bearable. "We've lived here for two years," said Akhi, the most talkative of the women. They worked shifts of eleven hours, and sometimes more, in a nearby garment factory in Dhaka, the capital of Bangladesh.

I had traveled to Dhaka to find where the jeans I was wearing came from, and the numerous other cheap garments labeled "Made in Bangladesh" on sale in stores across Europe and the United States. I anticipated finding the unacceptable side of the Bangladeshi garments industry, and discovering a place where sweatshop labor was even cheaper than in India. I was right on both counts. But I also discovered a strange flowering of female emancipation. These three young women, and hundreds of thousands more in this teeming city, were the first in countless generations of their families in conservative, rural, Muslim Bangladesh to have any sort of economic independence, any sort of personal rights beyond those allowed by their husbands. Life here was better than in the rice paddies and prawn farms. It was freedom and wealth, of a sort. And they were rather proud of it.

I had found the women by ducking beneath an overpass in the Mohakhali district of Dhaka, one of the city's many garment-manufacturing zones, then taking a dark lane down beside the railway tracks toward a food market and entering a warren of alleys through an unmarked door. There were eighty-four families packed in here, each occupying a single room. This was typical of where the seamstresses and occasional male coworkers of Dhaka's huge garment industry live.

Aisha, Akhi, Miriam, and the two other absent women each paid 500 taka a month for their room. Or $7.20. You wouldn't expect much for that money. But on wages of 1,660 taka a month, it was all they could afford. The overcrowding was severe. Men were showering in the alleys from hosepipes; children rushed about. The women got up early to cook breakfast and food to take to work. They had to because the four gas burners outside their room were shared by nine rooms, inhabited by forty people. "We all start work at the same time, eight a.m.," said Akhi, "so we have to start cooking at four a.m."

I could not visit their factory, but I saw others like it. Dark, dingy sweatshops up narrow flights of stairs, often with bars on the windows and blocked fire exits. The noise of sewing machines could be deafening, the managers intimidating, the drudgery unimaginable. When I met the three women, it was about eight in the evening. They had just gotten home after their eleven-hour day. One of them hemmed trousers, another sewed collars onto shirts, the third put in zip fasteners. Day in and day out. They didn't even get to swap jobs for the sake of variety. Their managers couldn't see the point of that.

"Officially we get one day off a week, but if there is extra work we have to carry on working," said Akhi. In Bangladesh, there is a legal maximum of fifty hours a month overtime, and employers are not allowed to require women to work after 8 p.m. But sometimes, she said, they worked all night and could clock up a 120-hour week. I had no way of checking these figures but no reason to dispute them. These were ordinary workers telling me their ordinary stories.

Who do they cut and sew and press shirts and jeans for? You and me, of course. The biggest customer at Aberdeen Garments, where two of the women worked, was H&M. Hennes & Mauritz is a Swedish company "known for their inexpensive and fashionable clothing offerings," as Wikipedia puts it. My daughter agrees. She buys a lot from H&M. And H&M buys around half of its cotton textiles from Bangladeshi garment makers. Most analysts believe the company is the biggest European customer in the whole of Bangladesh. So my daughter will very likely have something from Aberdeen Garments in her wardrobe—purchases that are part of my financial footprint, even if I don't wear them.

I asked these women what they thought about H&M, expecting an angry reply. But no. The garment workers constantly swap notes about working conditions, and H&M has a better than average reputation. The company pays more attention to their needs, they felt. So it was disturbing to hear claims that even H&M, for all its goodwill, was being hoodwinked.

The buyers—the brands' representatives in Bangladesh—make regular inspections of the factory, the women said. But they always inform the owners first. "Before they come, the managers come through the factory with megaphones. We are told to prepare the factory, to clean up. And they instruct us what to say about working hours and holidays and conditions. We have to lie about holidays especially." I asked about other brands. Of one, they said, "They are only concerned about the product." There was one factory where that company was the only customer. "That is the worst factory in Mohakhali."

Most of the four-thousand-plus garment factories in Dhaka have English names, uplifting names like Harvest Rich and Sunman Sweaters, Glory Garments and Niagara Textiles, Life Garments and the Abba Group, Imperial Knitting and Melody Garments and Fashion Club International. Some companies do their best, I am sure. But overall, this is a world of corner cutting, subterfuge, and threats. In recent times, the women said, buyers have threatened to withdraw business after reports of workers being sacked without reason. So the factory managers found a way round the problem. "They demanded that we sign a resignation letter. If they later decide to sack us for any reason, and if there is a protest about it, they will show the resignation letter to the buyer."

I heard dozens of similar stories, both scandalous and petty. One supplier to Gap, JCPenney, and Wal-Mart converted a storeroom into a daycare center when buyers came visiting, and told women to bring in their children for the day. Government inspections? Nobody had any experience of them. Not surprising, perhaps, since I was repeatedly told that most of the garment factories were owned by leading politicians and their friends and families.

Khorshed Alam writes reports on the companies for Western NGOs and retailers alike. He said most brands had little idea what was going on, and many did not really want to know. After a well-publicized scandal about alleged use of child labor, he investigated the Harvest Rich company. Its buyers included M&S, Tesco, H&M, Sara Lee, and Reebok. "Many of these brands had sent in social auditors to check conditions and interview staff, but when I read the audits they all said different things. It was as if they were writing about different places."

I talked to workers who had come to the Awaj Foundation, an NGO providing advice on labor issues. Some said that if they took a day off sick, they would be refused admittance to the factory and then fired for nonattendance. One was excluded for taking five days off work when her husband died. Others had no employment contracts at all. One man said he was sacked for refusing to sign a blank sheet of paper. Others for speaking up about conditions, or saying the wrong things to buyers. They reeled off the brand names that their employers contracted for: Wal-Mart and Gap, H&M and M&S, Sears and Asda.

But the women also told me about their former lives in rural Bangladesh. The three women in the Mohakhali slum all came from villages around Dhaka. Akhi had seven brothers and sisters. Back home there wasn't enough land, and certainly not enough work, to support so many. So the families sent their young women to find jobs in Dhaka. Out of their meager earnings, Aisha and Miriam, sisters-in-law, together sent home 4,000 taka a month (about $60). The alarming truth was that these women, for all their pitiful surroundings, were the rich ones in their families. And it confirmed what I heard in the prawn villages around Khulna—that a job in a garment factory in Dhaka was an aspiration.

Aisha and Akhi both had children back in their villages. They said they hoped to go back one day. But that seemed an idle hope. Most don't go back, and these women already seemed caught between two worlds. I noticed a cheap bag hanging on a hook at the back of the room. GUCCI was printed in large letters on the front. It was a fake, of course. But unlike their mothers and sisters back in the village, these women had heard of Gucci. They aspired. Even their tiny black-and-white TV showed them a world they badly wanted. What else was there to dream of when making clothes for Western consumers than of joining them?

But the economic gap between their world and ours is huge. In 2006, the Bangladesh government raised the minimum wage across the country for the first time in a decade. It was now 1,662 taka a month—a bit under a dollar a day. Or 10 cents an hour. No wonder the likes of M&S are moving here. Why pay 35 or even 70 cents an hour when you can pay 10 cents instead?

But even 10 cents was proving too much for some Dhaka garment makers. Many were simply delaying monthly wages to maintain cash flow. I met Jakir, who had just been sacked from his job ironing and packing T-shirts for a garment factory that sold to NKD, a German discount clothes company. The factory was owned, he said, by a well-connected former captain of the Bangladesh cricket team. Jakir's crime was to lead a delegation complaining that wages for February would not be paid till March 30, and February overtime not till mid-April. He said there was a lot of overtime pay due because the workers were routinely forced to work a fourteen-hour day and sometimes nineteen hours "because of the contract pressure." I later checked NKD's website. Its T-shirts retailed at about $7.50.

Child labor is largely banished now from the factories that Western buyers know about. But it persists on a small scale among the subcontractors that the factories call on to meet customers' deadlines. During my visit to Dhaka, I saw a locked-up building right next to an NGO in Shyamoli, a middle-class suburb. My host said it was a garment factory that operated from time to time. The workers appeared to be children. He often heard their voices amid the sound of machinery, in the middle of the night. He had never been inside, but materials for garment making often came in and out of the building.

He seemed nervous about intervening, and didn't know who the children worked for.

Nazma Akter, the founder of Awaj, said some buyers were sympathetic about the obvious exploitation of both children and adult workers, but "in the end they are not serious. Price comes first. Every time the buyers make another order, they want lower prices. First it is the $10 T-shirt, then the $6 T-shirt. That is unethical." The system, she said, is run "so that everyone down the line is well enough off to have a nice house and a car—except for the women who make the clothes. Sometimes I even feel sympathy with Bangladeshi manufacturers."

Soon after, I spoke to a top manager at one of the main garment makers. She said, "These big-brand companies have corporate social responsibility departments, but the people who make the orders don't talk to them. We see it here all the time. The CSR people come in and stipulate basic standards. Then the next day the buyers come in and drive down prices and bring forward deadlines. These two things are usually completely incompatible. While the buyers are in charge, all this talk from the CSR people is just corporate window-dressing, however well-meaning the people involved might be."

The truth is that retailers talk about little else but price, and how they are moving from Sri Lanka or India or Mauritius to Bangladesh or Cambodia or China to take advantage of lower wages. The only limitation seems to have been the quota systems set up by European and North American governments to protect their own domestic industries. But the quotas are now being abolished. And the "race to the bottom" in the mass marketing of cheap clothes is intensifying.

Nazma dismissed Western hand wringing, and said she thought that it would be pressure from within that would ultimately clean up the industry. But she pleaded with customers in the West not to respond with boycotts. The jobs, poor as many were, empowered women. "Women are becoming an economic force. This is the first time they have had jobs outside the home. They are independent decision makers. They can come and go; nobody stops them. They are the economic backbone of the Bangladesh economy now." Western consumers, she said, should be demanding better conditions for the women of Dhaka, and above all should be willing to pay the higher prices involved. And retailers should stop competing on price. But please, she said, "don't stop buying."

Most of the garment workers I spoke to agreed. Like many other workers I met on my journeys for this book, they had a good sense of their worth, and how they were being let down by the obsession of Western consumers with ever lower prices. Shuktara, who came to Dhaka eight years ago, told me, "Our pay in a month is less than the price that you pay for a pair of jeans." That was about right. Her pay works out at $30 a month. Sitting next to her, Nosrul, who came to Dhaka twelve years ago, chimed in. "In our factory, there are between eighty and ninety workers. In an hour we make a hundred and twenty pairs of trousers." So it takes a bit less than an hour for one worker to make the trousers that will eventually be sold for the price of her salary for a whole month.

Bitter? If so, Nosrul hid it well. "We just want the customers in the West to know that we would like them to pay more for their clothes so that maybe we will get paid more too." I can't imagine myself being so sanguine.

Caught up in the human stories, I had lost track of the story of my jeans. I had bought a pair of Blue Horizon jeans for the equivalent of about $18 in Marks & Spencer a few months before. Made in Bangladesh. M&S bought them from a company called, coincidentally, Spencer's Apparel, part of the large Medlar Group. The fabric that made up the jeans came from the Nassa Group, which was Wal-Mart's "International Supplier of the Year" in 2002. But M&S didn't want to give me an introduction to the companies, and without that the companies weren't interested in talking to me. So I asked Khorshed about them. They weren't the worst, he said, but not the best either. Just one of the pack in the race to the bottom.

I wanted to follow where my jeans came from back to the cotton fields. And here came a surprise. Bangladesh's largest garment makers want to cement their position in the market by making their own fabric. In pursuit of that goal, Bangladesh has become the largest consumer of cotton from the Central Asian state of Uzbekistan.

White Gold   13
My T-Shirt, Slave Labor,
and the Death of the Aral Sea

It is an old story, but with an unpalatable new twist. One we should
be ashamed of. The world knows all about how, under Josef Stalin,
the Soviet Union harnessed the rivers that flow into the Aral Sea, extracting all of their water to irrigate fields of cotton. It knows how
the cotton fields that clothed the Soviet empire were worked by what
amounted to slave labor, with millions of people, including schoolchildren, teachers, and doctors, press-ganged into harvesting the
"white gold." And it knows that this social holocaust was accompanied by an ecological holocaust as the Aral Sea, the fourth-largest inland sea in the world, dried out.

All that was supposed to stop when the Soviet Union collapsed
in 1991. Four years after Moscow's retreat, I joined officials from
the World Bank and UN Development Programme at an international conference in Uzbekistan, the biggest cotton producer, where,
together with the governments of the region, they agreed to the
creation of the International Fund for Saving the Aral Sea. Schoolchildren were invited to the meeting to spray the delegates with flowers and declare their happiness that the sea was going to return—and
that their days of forced labor were over. Little did they know. When
I returned a decade later, not only had the sea not returned, but the
plight of the people and the entire local environment had deteriorated. The enforced labor continued, and even less water reached the
sea than before. Wherever the money from the international fund
went, it did not go into reviving the sea. And now we are to blame.
For the cotton that once clothed the Red Army is now clothing us.
Today it is found on the shelves of most retail stores—and on our

backs. Yet retailers who pride themselves on their ethical and green credentials—and who can give you precise details about the source of their "fair-trade" cotton—profess to be blind to this tyranny.

Since the collapse of the Soviet Union, the assault on nature and the citizens of Uzbekistan has intensified. Islam Karimov, the former strongman of Soviet Uzbekistan, remains in place, and seems to have been emboldened by the departure of his Moscow masters. Most years, no water at all makes it down the Amu Darya, which once carried more water than the Nile. Cotton still employs 40 percent of the Uzbek workforce, including hundreds of thousands of schoolchildren. And complaints about human rights abuses have intensified. One story in the London *Sunday Times* began: "When Fazliddin Akhrorov was ordered by officials in the central Asian republic of Uzbekistan to give up his studies and pick cotton for three months, he at first refused. Threatened with expulsion from his institute, he was taken away. Within three weeks the healthy sixteen-year-old was dead."

Uzbekistan is the world's second-largest cotton exporter, after the United States. Most of its 880,000 tons a year goes south by train to the Iranian port of Bandar Abbas, and then by ship to south and east Asia. Of the 440,000 million tons of cotton imported by Bangladesh, half comes from Uzbekistan. That makes Europe and North America, the biggest consumers of clothes from Bangladesh, the main destination for Uzbek cotton.

We are rightly concerned about the sweatshops of Bangladesh and elsewhere. But we seem profoundly indifferent to the people who grow and harvest the cotton. I wanted to put that right. I don't know for sure that the cotton in my $18 jeans came from Uzbekistan, but I can surmise that it probably did.

I traveled the length of Uzbekistan, from Tashkent to the Aral Sea. What struck me most was that the old Soviet system persists, but is living on borrowed time. The nation is straining ever harder to deliver its cotton harvest, but the actual output is less every year. The reason is that the infrastructure is decaying, and the environment is deteriorating. Evidence of the decay since the departure of Commissar Cotton appears around every corner. Canals are filled with weeds, sluice gates are broken, and concrete water channels leak into the ground.

The Russians prided themselves on leveling the fields to prevent waterlogging. Few farms do that anymore. So the water builds up, leaving behind salt in the soil. Salt is toxic to cotton, so every winter the fields have to be flushed clean. These days, around half of all the water taken from the rivers is used not to irrigate crops but to flush the salt from soils before the spring planting season. Meanwhile, with drainage systems overwhelmed, huge lakes of salty water are forming in the deserts. I spotted one that took an hour to drive past. The Aral Sea has not so much been dried out as dissipated into a thousand desert sumps.

But I found the worst consequences of the country's continued addiction to growing cotton at the end of my journey, in Karakalpakstan. This is the Uzbek province that once bordered the Aral Sea, before the tide went out. It had the country's seaside resorts and a huge fishing fleet operating out of Muynak. Now, you can walk north for more than 60 miles from Muynak before you find the sea. Not that anybody does; it is a new and unexplored desert.

Karakalpakstan was once a green and pleasant land, the delta of the Amu Darya, once called the Oxus. Even twenty years ago, it still grew large amounts of cotton. But now the water is all taken upstream. Little water reaches the delta, and virtually none makes it to the sea. The province is turning to desert. On the delta, half of the old cotton fields have been abandoned. Old state farms have gone bankrupt and found no buyers. Tamarisk is returning to the fields, and the population is migrating, or going back to the old ways—herding cattle.

Nothing I saw here was new. Everything was old Soviet issue. The cars were old Ladas. Soviet automatic cotton-picking machines lay rusting in scrap yards. More work for the children. I came across the site of a large Soviet tractor-part factory in the town of Kanlikov. Today it is empty and the railway sidings that once brought raw materials are covered in weeds. Each year, locals said, there are fewer tractors and more donkeys.

The Uzbeks were told that, after 1991, capitalism would bring investment and efficiency. The truth seems to be the opposite. There is no investment; efficiency has deteriorated; concern for the environment appears to be even less than before. The only surviving Soviet skill seems to lie in propaganda. During my visit, the evening news announced that Karakalpakstan had met its production target for the first time since 1993. What the newsreader never mentioned was that the target had been halved in that time.

Uzbekistan takes from nature more water per head of its population than any other nation, except for its smaller cotton-growing ex-Soviet neighbor, Turkmenistan. Yet I visited communities in Karakalpakstan where people had nothing to drink except water from irrigation canals. And that water was the uncleansed salty effluent from upstream farms. The water—indeed the entire environment of Karakalpakstan—is laced with salt. It is in the water, the soil, the food, and the air. As a result, anemia has become endemic, infant mortality is soaring, and cancers of the esophagus are the highest in the world. Since Soviet times, life expectancy has fallen from sixty-four to fifty-one. Local gynecologist Oral Ataniyazova told me that in many villages on the delta the farm water is so salty that cow's milk added to tea curdled. And sometimes babies will not take their mother's milk from the breast because it too is too salty.

Without the moderating influence of the sea, the climate has changed. Summers are hotter but shorter. Winters are colder and longer. People often harvest the last cotton in November, desperately trying to reach their quotas with frozen fingers on frosty mornings. Rainfall has declined and the region is increasingly ravaged by dust storms. An estimated 77 million tons of dust from the exposed seabed blows across the land each year. It carries a cocktail of salt and pesticides brought to the sea in past decades in drainage water. Those pesticides include DDT and lindane, which are banned in Western countries but still fill the air here.

"The entire population of Karakalpakstan has been chronically exposed to salt and farm chemicals," Oral told me. "We have researched the literature but can find nowhere like this anywhere in the world." At a meeting of farmers in the delta town of Chumbai, one weather-beaten man, the manager of a state cotton farm, looked at me gravely and said, "We are all affected. There is a lack of good blood. All the women have it. Many children are born deformed here. My daughter and son were both in hospital for six months."

Cotton has become a curse to Uzbekistan. A curse maintained by Western markets for cheap cotton clothes. "We don't need cotton; we need fruit and vegetables," Oral said. "Then both the sea and the people could recover." Maybe it will happen. The cotton industry is probably dying. Everywhere I went in Karakalpakstan, people said that when they had spare cash they bought cattle. That is what they did in the old days before cotton. The herds are growing. As I turned back from Muynak and headed for Tashkent, the roads that once hummed with tractors from the cotton farms were full of cattle coming home from grazing on the abandoned fields. The people may be giving up on cotton, but the government is not. At least not yet. Instead, the country's elite is trying to get rich by it. The nation's cotton exports are all controlled by a series of exporting agencies, with names like Uzimterimpex and Uzmarkazimpex and Uzprommashimpeks, each of which shipped more than 275,000 tons of cotton in 2004. The agencies are privately owned and controlled by state officials and people close to the government. According to former British ambassador Craig Murray, the agencies rip off the farmers, who receive only a thirtieth as much as similar farms in neighboring Kazakhstan.

The International Crisis Group discovered that the agencies often work through offshore companies registered in the British Virgin Islands or Cyprus, laundering cash from their cotton trades with the big global brokers into underworld accounts. At any event, the cotton profits seem to translate into shiny new buildings in Tashkent and shiny new Western cars for top officials in the agencies.

If I can put this simply, the obsession of many people (yes, including me) with ever-cheaper jeans is not just helping to sustain the dreadful conditions in Bangladeshi sweatshops, it is also helping enslave Uzbeks, desecrate their land, and finish the emptying of the Aral Sea. And while the Bangladeshi sweatshops do have the saving grace of helping rural women take the first step on a long road to emancipation, I can see no similar justification for propping up the Uzbek cotton farms. They are bad, bad, bad.

I learned all this during a weeklong visit to Uzbekistan, escorted by locally based researchers from an international agricultural research institute. Their status gave the scientists some independence and safety from intimidation by the government there. But it seems that few of the cotton traders or retailers who buy up the country's cotton harvest have taken the trouble to make a similar journey. Thomas Reinhart, who runs the cotton trader Paul Reinhart, told the London *Sunday Times* he had never heard of the use of child labor in the region. "We buy our cotton from government agencies and don't know what happens out in the fields." Well, I reckon he damned well should. It isn't hard. And he has a large staff in the country to figure it out.

Part Four

The Chinese Dragon

Computing Power   14
Mice, Motherboards, and
the New Emperors of Suzhou

You will never have heard of Sammy Cheng. I hadn't. But we are all in touch with him. Probably more of us touch his products, and do it more often, than any other physical object coming from one place. For Sammy—an ebullient, shirt-sleeved raconteur in the buttoned-down world of Chinese businessmen—runs the factory that makes two-thirds of the world's computer mice. It also makes a third of the world's webcams; has a virtual monopoly in trackballs, the upside-down alternative to the mouse; and claims a half share in the world-wide market for remote controls. And how many of those do you have around the house?

Sammy runs the main manufacturing base for the Swiss computer peripherals company Logitech. Its six thousand employees work on the fringes of Suzhou, one of China's most ancient and beloved towns, famous for its Confucian gardens and waterways. But Suzhou is also just fifty minutes by train from the megacity of Shanghai, and is in the middle of what is fast becoming the world's largest urban area, stretching from Shanghai across the delta of the Yangtze River. And if you wanted to bring the world's computer manufacturing industry to a halt, shutting down the power station at Suzhou would do it. For on Suzhou's outskirts, now dwarfing the ancient town, are two giant, newly built industrial cities. And besides making Logitech's indispensable computer peripherals, Suzhou hosts factories that make a quarter of the world's laptop computers, a tenth of its scanners, and nearly half of all its PC motherboards.

I went to Suzhou to see where my computer world came from, but also found myself plugged into Sammy's world. Sammy is Taiwanese—the son of an air force colonel and an electronic communications expert who both fled mainland China in 1949 when the Communists took over. Now he is back, and so are large numbers of his engineering friends. Virtually all the companies that have made China the heart of world computer manufacturing have their headquarters in Taiwan. And Sammy was bursting to introduce me to his compatriots and to give me a snapshot of their takeover. But first, my mouse.

Logitech is Swiss owned but with headquarters in California's Silicon Valley. It was one of the Suzhou pioneers, establishing its main manufacturing base there more than a decade ago. That is almost medieval times in the footloose world of computer electronics. Now it churns out 70 million mice a year—under its own brand name and for almost everybody else as well. I watched them stamping the names of Dell and Acer and Hewlett-Packard and many more onto the mice as they came off the production lines, one every three seconds. And Sammy was gearing up for the release of the world's first mass-produced 3-D mouse, for Christmas 2007. "Our philosophy is to have more than fifty percent of any market," he said with calculated immodesty. "That's why we have had double-digit growth here in both output and profits for the last thirty-four quarters."

Keeping the mice flowing is a high-wire act. The mouse may seem like a simple peripheral, but Sammy has four hundred suppliers, and if any one of them goes down, he is in trouble. Luckily most of the suppliers are local, with 90 percent of the manufacturing within 30 miles. For instance, his capacitors, the tiny energy storage devices needed in most electronics equipment, are made by the billion by Yageo, the Taiwanese market leader, just across town. But the two most high-value components for bestselling mice come from abroad —chips from Motorola's plant in Malaysia and optical sensors from Agilent in the United States.

Suzhou is a boomtown. Driving down the main street, Sammy said, "There were no cars here three years ago, and look at it now." But real wealth is rare. While a handful of senior Taiwanese managers earn Western salaries, their more junior Chinese managers are on salaries of about $26,000 a year, and the five thousand or so production-line operators earn less than $2,000. Once a paragon of so-cialist equality, China now has the biggest income gap between rich and poor anywhere in the world.

The eighteen-year-old girls who run the assembly lines are de-livered by agents from the country's poorer provinces, like Anhui and Wuhan, which is where Sammy's parents came from. They are part of the estimated 100 million members of the greatest mobile la-bor force of all time, and among the most compliant. Most live in cramped Logitech dormitories, six to eight to a room. For recre-ation—between shifts where they make trackballs and webcams for the world's computerized youth—the girls have basketball and table tennis.

Touring the plant, I waited in vain for one of them to look up from their work. Just once, a girl with a plump face caught my eye. It was partly flirtatious, but mainly bravado. The girl knew she was alone in acknowledging the rare foreigner visitor. From her glance, I imagined she knew she was being exploited, but also that she had found freedom of a sort from the stifling world of the Chinese peas-antry. Freedom, at any rate, to glance at foreigners.

Logitech may dominate the market in sexy computer peripher-als, but the real heart of a computer is the primary circuit board, or motherboard, where the microprocessor and central memory are housed. The world's largest manufacturing center for motherboards is virtually across the street from Logitech. Asustek makes almost half the world's motherboards, 60 million a year on a campus of breathtaking size that employs eighty-five thousand people. "It's not a company; it's a whole town, an entire society," said Sammy envi-ously, as we drove up to one of its six heavily manned gates. There are no factories in England today one-tenth the size of Asustek's opera-tion here in the ancient city of Suzhou.

You may not have heard of Asustek. It is one of the many "ODM" companies that rule here. ODM stands for original design and man-ufacture. Besides making motherboards, they design and manufac-ture the computers that are sold under Western brand names. They are the key to Taiwanese success in the world computer business. The first ODMs were Taiwanese companies manufacturing the first laptop computers, and the business model has spread to dominate the rest of the industry. Buy a Hewlett-Packard or an IBM or an Acer, and the chances are that Asustek has not only made the motherboard, but also assembled, packaged, and shipped the entire computer from here. I am writing this book on an Acer, almost certainly from here. Asustek also makes the "bare bones" of Dell computers, though Dell finishes them off, customizing them to order at its own plant. And if Asustek didn't make your computer, then probably one of its ODM rivals, like Quanta or Compal, will have.

Sammy introduced me to his friend David Chen, another cheery Taiwanese and Asustek's procurement vice president. The two men had worked together years ago for the Dutch electronics company Philips. These days, David runs a logistical nightmare that puts even Sammy's work in the shade. He has to ensure that components from 1,800 different suppliers always turn up in time for assembly— including mice from Sammy.

The Asustek complex contains six separate factories. Factory six, the largest, employs twenty thousand people. I had only an after- noon, so I toured plant number one. It has sixty-eight production lines, making motherboards for all the major PC brands, including the top four: Dell, HP, Acer, and IBM. On the lines, the basic task is to plug and solder resistors, capacitors, and other electrical compo- nents onto the boards. Some of this work is mechanized, using big machines from the German manufacturer Siemens that keep going twenty-four hours a day. But many of the larger components are in- serted by hand.

As at Logitech, this is the work of China's most marketable re- source: bored, methodical, nimble-fingered eighteen-year-old girls fresh out of school. Besides plugging in the components, some girls do hand-welding. They have simple textile face masks, and lead is no longer used in the solder. But the pungent solder fumes spread down the production line. David said many of the fiddly repetitive tasks be- ing carried out by the girls were "hard to do by machine." Managers at other assembly plants told me there were machines up to the job, but girls were cheaper. At any rate, said David, "we only use women for these jobs; they have better fingers."

Asustek has a reputation as the lowest-cost ODM. It works its machines and its staff hard. The basic shift is eight hours, and operators work a six-day week. But, rather than running three shifts, Asustek has two shifts plus routine overtime. That may be as little as two hours in the low season, but it stretches to four hours in the high season, in the run-up to Christmas, when there will be operators on the production lines twenty-four hours a day. It might be worth bearing in mind next Christmas that a computer bought around then may have been made by people in the twelfth hour of their shifts.

As in many low-wage industries, the workers depend on overtime. Sammy told me, "People leave if they don't get overtime. It is the only way they can save." I asked how long the operators at Asustek stayed.

"We have a turnover of ten percent," said David.

"Ten percent a year—that's amazingly good," I said.

"No, not a year, a month." So most people leave after less than a year. Despite inducements like free train tickets, huge numbers of the girls never return after going home for the Chinese New Year.

Wasn't he worried that women were so keen to leave his employ? Sammy intervened: "These jobs are one of the starting points for rural girls setting out on a career. When they leave, they get other work in the city, restaurants for instance." In which case, why do so many leave at New Year? While some girls stay in the city to find other jobs and send money home to their families, many do not. The image of tens of thousands of girls going home to their families and never returning told its own story. A story of young women lost and disillusioned in the vast regimented confines of a factory like Asustek. Enslaved, we should remember, to the demands of Western brands and consumers. My Acer PC, which I confess I bought largely because it was cheap, is a critical part of that process.

David was more concerned about the practicalities than the ethics. Asustek has to find and train seven thousand new operators every month. He has recruiting agents touring schools in almost every province in China. Sammy likened it to army recruitment. And every morning, eight hundred buses show up at the plant bringing in workers from dormitories and digs across town. Just getting food for the canteens is a problem, David said. "Before last New Year, we tried to do a special meal, but nobody could supply forty thousand chicken legs, so we settled for cake." Let them eat cake. It left me wondering what they ate the rest of the time.

Suzhou is expanding in two directions: west to Lake Taihu, China's third-largest lake, and east toward Shanghai. The western park, known as the Suzhou New District, is cheap and basic. Rents are low and this is where the companies with large payrolls like Asustek set up. The eastern Suzhou Industrial Park is more prestigious, but higher rent. It looks rather like Singapore, all glass and trees and wide boulevards—probably because it was designed jointly with the government of Singapore. This is where the upscale apartment blocks are congregating, where the city authorities have recently relocated, and where education campuses and science parks are being installed in the hope of attracting more high-tech, R&D-based firms, like software companies and nanotechnologists.

The city claims to have six billion-dollar investment projects on the go out here. It is a maze of foreign brands: buying, selling, and manufacturing. Tesco and Siemens, BOC Gas and Knorr, Pfizer and Sony, Epson and Sanyo, Alcan and Bosch and Wal-Mart. If China escapes from the low-wage sweatshop origins of its industrial revolution to become a fully fledged consumer society, then this is how it will happen.

Keep going on the expressway east of Suzhou Industrial Park and you get to the Kunshan export-processing zone. Here another Taiwanese ODM, Compal, makes almost a tenth of all the world's laptop computers. It produces for Dell, IBM, Hewlett-Packard, Acer, and many other brands. My trusty Toshiba, companion on many a trip, came from here. "We sell them webcam modules," said Sammy as we headed in to meet another old mate, Compal vice president Mage Chang.

Compared to Asustek, this place is modest. Just twenty-two thousand employees. Compal was one of the earliest laptop manufacturers, and now this plant turns out about a thousand machines a year for every one of its workers. That works out at a retail value of about $1 million a head, of which each employee ultimately sees about $600. But Mage told me margins were tight, with the big brands demanding ever-lower prices.

The immediate competition was literally over the back fence, where another ODM called Wistron, a spinoff from Acer, employed a workforce almost the same size as Compal's to make almost as many laptops. Rivalry was intense. Compal was, at the time of my visit, the world's number two notebook maker, behind Quanta, whose main operation is in Shanghai. But it was only a whisker ahead of Wistron. Mage told me he regularly looked over the fence to count the number of container trucks leaving his rival's yard. He might have been joking, but I don't think so. Compal is another major supplier of bare-bones computers to Dell, which turns them into complete custom-built machines at its assembly plant 435 miles to the south in the port town of Xiamen. I followed the bare bones there. I was thinking of buying a Dell notebook next time. Dell famously claims to ship completed computers, made to order, within twenty-four hours. And it seems to be true.

You might imagine that meant holding lots of stock. But not so. Dell buys $16 billion worth of bare-bones PCs and components from Chinese companies every year—a spend exceeded only by Wal-Mart. Yet at its Xiamen plant, it holds no stock at all. Instead, it instructs its suppliers, like Asustek and Compal, to bring supplies to the city and hold them in warehouses close by. Every two hours, Dell sends out new orders for parts, which it expects to be delivered to its premises within thirty minutes—"just in time" for assembly.

I watched. Assembly begins, rather like a cafeteria, with a large pile of pink plastic trays. Each receives a piece of paper detailing the individual order—even a Dell production line is not paperless. Then components are loaded onto the tray, and the full tray is sent for assembly. A few hours later, after software is installed and tests run, the completed computer is packed up into a box at the other end of the building, addressed, and shunted through a hole in the wall straight into a container for dispatch.

Interestingly, Dell does not use a conventional production line, in which workers pick a product off the conveyor, do their single task, and pass it on. Instead it has one hundred production "cells" in which single workers perform most of the assembly in less than ten minutes. They have to be better trained than on a conventional line, but the job is more interesting. And they are slightly better paid, at a bit over $2,000 a year. Staff turnover, Dell said, is lower as a result. Oh, and guess what. At those pay rates, they appoint as many men as women to the operator jobs.

China makes 80 percent of the world's PCs and laptops. But the brains behind this remarkable dominance are Taiwanese: men like Sammy and David and Mage and bosses back in Taipei. Whatever the label may say, it is they who design and manufacture your computer. Today eight out of China's top ten exporters are Taiwanese ODMs, supplying big Western electronics brands. They have $12 billion invested in Suzhou alone.

Many of the Taiwanese engineers and entrepreneurs who run these companies worked first for Western electronics companies in Taiwan, like RCA and Philips. As Sammy tells it, as each of these companies shut down, the thousands of engineers they had trained set up their own companies: Asustek, Compal, Acer, and its offshoot Wistron, Flextronics and Hon Hai, the owner of Foxconn, which manufactures iPods for Apple. The companies moved to mainland China to take advantage of low wages, but their top management is still dominated by Taiwanese.

As many as a million Taiwanese electronics managers now work on the mainland. They form a kind of economic occupying force that must be worrying for China's leaders in Beijing. And despite superficial rivalry, Sammy says there is a common bond among the ODMs. "The top management were colleagues and schoolmates. We all served in the Taiwanese army. It's a team culture," he says. "If we need parts, we will call each other up."

Sammy has done well. Not content with homes in Taiwan and Suzhou, he has set up his family in Canada, where his wife lives in Vancouver and his children are at universities in Toronto. He visits them en route to his monthly reporting sessions at Logitech's headquarters in California. His escape from China is complete. But others are left behind. And some are victims of the helter-skelter economic growth.

Across the road from Logitech's plant in Suzhou, there are row upon row of mean twelve-story apartment blocks. They accommodate the thousands of farmers whose land has been paved over to make way for the new factories. The farmers are allocated menial jobs in the city. Dozens of villages have been wiped off the map already. And the municipal authorities are in the process of extending the Suzhou New District by more than 75 square miles, an area slightly larger than Washington, D.C. That will take it down to the beautiful Lake Taihu. But already the pollution from the factories is playing havoc with the lake's ecosystem. It has triggered a recurrent plague of algae. In 2007, the government declared the lake a natural disaster and hastily vetoed plans for the construction of a new prestige residential area on its shores.

But the industrial development continues unabated. At the end of my day at Logitech, Sammy drove me down to the lake, where, whatever the new edicts, they are building golf courses and condominiums. On the way, we drove through the old town of Dongwu. It is famous for its silk embroidery, but is rapidly becoming a building site. And judging by the maps drawn up by the municipal authorities, it too is scheduled for elimination.

Zhangjiagang   15
The World Capital of
Rain-Forest Destruction

Zhangjiagang is not far from Suzhou. It is a major port on the Yangtze River, and imports more tropical hardwood timber than any other port in the world. Day and night, the ships cross the South China Sea and come up the foggy channel of the Yangtze, past the megacity of Shanghai and wharfs where wastepaper from Europe and North America is imported, before unloading their cargoes of okoume from West Africa, or meranti from Borneo, or teak from Burma, or merbau from West Papua, or greenheart from Guyana, or bintangor from Papua New Guinea.

More than 1,600 acre-feet are unloaded here annually. Zhangjiagang is the world capital of rain-forest destruction. The place where, as they joke in China, the forests of Borneo are laid out end to end and turned into flooring for new apartments in Shanghai and Beijing. Except that in truth as much of the timber is exported again as stays in China. For in the past decade the "workshop of the world" has also become the floorboard fabricator, plywood maker, and joiner of the world.

China is far and away the largest buyer of timber on the planet. It is also the world's leading exporter of plywood, wood flooring, and furniture, where it has more than a third of international trade. The Tropical Forest Trust, which works with European retailers to green their timber supplies, estimates that half of all the internationally traded tropical hardwood passes through China, and two-thirds of that passes through Zhangjiagang.

I don't buy too much wood personally. My last purchase of furniture was a solid beech desk made in North Yorkshire from timber probably blown down by the 1987 hurricane. My computer sits on it. It will see me out. But I am guiltily aware that I also have a wooden table and bench out in the garden. They were probably made in China or Vietnam, from timber hacked down who knows where. And there are the kitchen units, which have what looks like a tropical hardwood veneer. Gulp. Campaigners say that if I don't know where timber like this comes from, I should assume that it was logged illegally. I am in receipt of stolen goods—stolen from the rain forests.

At Zhangjiagang, local merchants happily show you round the vast timber yards, where there are street signs to prevent you getting lost. Many of the merchants have no calling cards, only cell phone numbers, which they scrawl onto the logs with a marker pen. Xiufang Sun, a market analyst for the nonprofit research group Forest Trends in Beijing, knows the port well. She says there are about 140 importers and dealers working full time there. Most have no idea where the logs they buy come from, she says, though they must know they have often been illegally felled. But so far as they are concerned, that is someone else's responsibility. The wood was dead on arrival. They simply want to sell as fast as they can, to make room on the dockside for the next load coming in on the next tide. And they have plenty of clients, as the rash of five-star hotels in the port underscores.

Most of the timber coming out of most of the countries from which Zhangjiagang imports is reckoned by reputable international agencies to be logged illegally. Sometimes this may be on a technicality. But often it is sheer brazen corruption and transgressing of basic laws designed to protect the forests and their inhabitants.

A short way up a delta creek from Zhangjiagang is a small town called Nanxun. It is not on most maps, but it calls itself the hardwood flooring capital of the world. With reason. In the past half decade it has gone from a small, local woodworking center to the largest producer of wood flooring in a country that exports half the world's flooring. It has some two hundred sawmills and five hundred floorboard manufacturers. They are lined up along the creek. Here, planks from Zhangjiagang are cut, planed, and varnished to make flooring for the world.

Much of the flooring is made of merbau, a tropical hardwood from the island of New Guinea prized for its strength and durability. It also makes great outdoor furniture and decking. The London-based Environmental Investigation Agency (EIA), which has tracked the merbau trade in detail, says one log of merbau from Indonesia's New Guinea province of West Papua is imported to China every minute of every working day.

One of the largest manufacturers of merbau flooring in the town is Zhejiang Fangyuan Wood, which produces 2 million square meters of flooring a year. Half of this, it says, is destined for export. This merbau trade is largely illegal. Not in Chinese law, let me make this absolutely clear, but in almost everybody else's. The EIA calls it "the world's biggest timber-smuggling racket." And it says China is "the largest buyer of stolen timber in the world."

China's emergence as a timber superpower is recent. Only a decade ago, it was virtually self-sufficient in wood. Then two things happened. First, in 1998 there were major floods on the Yangtze River. More than 2,500 lives were lost and millions were left homeless. The government blamed the floods on deforestation in the river's headwaters in provinces like Sichuan, and it swiftly banned all felling of the country's surviving natural forests. The ban was 100 percent effective. Four years later, I was in the mountains of Sichuan, north of Chengdu. Locals told me the roads there used to be bumper to bumper with trucks carrying logs. But now they were empty, except for the occasional bicycle. Huge timber-processing works stood empty.

But at the same time, China's wood products industry was taking off. The country erupted like a typhoon onto the world timber market. Within four years of the logging ban, it had gone from nowhere to importing nearly 13,000 acre-feet of logs. It was the world's largest importer, unseating Japan, which had held the position for several decades. The majority of that timber was softwood from the Russian far east. But the country had also become the biggest consumer of tropical hardwoods. Today it buys 70 percent of all the hardwood leaving Southeast Asia, and increasing amounts from central Africa.

First, China tapped Indonesia, especially the rain forests in Borneo and Sumatra. I was in Borneo in 1999 and watched the process firsthand. Every week, thousands of laborers arrived on ferries from Java, where the country's economic meltdown had put them out of work. They were taken from the ports by truck to remote forests, handed chainsaws and given $6 a day to cut down the trees. Groups of up to two hundred men were living rough in the forests, felling trees. I saw several teams of Javanese men jammed in trucks driving up and down the new trans-Kalimantan highway.

In places the first timber they cut was used to make crude railways across the thick peat swamps. Only then could they bring out the timber in quantity. The pace of deforestation since has been breathtaking. One of the largest rain forests on the planet, covering an area larger than England, has been stripped bare in the past decade. At least half of the logging was illegal. A great deal of the timber went to China. Is some of it in my garden bench?

The Indonesian government has recently clamped down on the export of logs. It is trying to protect what is left of the country's forests. But it doesn't have the authority of its Chinese counterpart. The military is strongly involved in logging and is a law unto itself. Many generals and their cronies have become rich. Merbau is one of the most valuable timbers in the Indonesian forests, especially in West Papua, a remote province ruled by the military, where loggers have moved as the pickings become thinner in Borneo. The EIA says several shiploads of merbau leave its waters every month, headed for Zhangjiagang. Much of this arrives with documentation saying it came from Malaysia, from where log exports remain legal.

The Indonesian crackdown has pushed up prices. A cubic meter of merbau can now cost $500. But with trade heating up and the logs getting expensive, Papua New Guinea has taken over from Indonesia as the timber pirates' forest of choice. Shiver my timbers, indeed. Papua New Guinea is now China's biggest source of tropical hardwood. And the place, some say, where the immorality of the trade is mostly clearly on display. At remote harbors one of the poorest countries in the world is emptying its forests of merbau and other valuable tropical hardwoods like bintangor, which is a favorite veneer for plywoods. Four out of every five logs that leave these shores end up in China. About 80 percent of Papua New Guinea is still covered in forest, but more than half is licensed for logging. Current estimates are that the country's forests will be logged out in fifteen years.

The logging of Papua is now a huge business. Timber is the country's main export earner. Fortunes are being made in this former South Pacific backwater. And about 40 percent of the logging business in Papua New Guinea is run by the Malaysian firm Rimbunan Hijau. This thirty-year-old company is owned by Tiong Hiew King and his Chinese-Malay family. Tiong has been ranked as Southeast Asia's twentieth-richest man. Today, his company is one of the world's largest loggers and timber traders, with subsidiaries in the Solomon Islands, Cameroon (where it has more than 345,000 acres of logging concessions), Gabon (135,000 acres), and Equatorial Guinea (80,000 acres). The company is noted for its intensive logging methods.

Tiong has established what many see as a virtual stranglehold on Papua New Guinea's logging industry. Industry analysts say Rimbunan Hijau and its network of subsidiaries and related companies own twelve of the top thirty concessions and are responsible for over half of the more than 1,600 acre-feet of logs exported each year. The Tiong empire also has the country's largest sawmill, at Kamusie in the remote western province, and its only veneer mill, at nearby Panakawa. And it owns one of the country's two major newspapers, the *Nation*, as well as having interests in fisheries and shipping, insurance and mining, finance and IT, retailing and palm oil plantations.

While there is plenty of legal logging, the World Bank estimates that 70 percent of the timber from Papua New Guinea is illegally logged. An official review of forestry allocations in 2003 looked at two controversial concessions called Wawoi Guavi and Vailala, and concluded that "the time has come for a full investigation into the affairs of these companies. They should be compelled to account for their actions."

The government in Port Moresby insists that the logging operations are fully licensed and that forestry laws ensure "sustainable" operations. It says independent audits for the government confirm this. However, Forest Trends examined sixty-three of these audits and concluded that "the overwhelming majority of commercial logging operations are illegal," with logging rates far in excess of any that could be sustainable, and few benefits accruing to the locals. Forest Trends says there is "corruption at the highest level of government and throughout the bureaucracy. Conglomerates with logging and export interests are reported to provide funds for political parties and individual politicians ... In return for funds, companies are able to buy the right to log particular concessions and they are virtually exempt from the rule of law."

We are part of this. The link from the forests of Papua New Guinea to the timber yards and workshops of China is clear. But so too is the onward path of products to the furniture, flooring, and DIY stores of North America and Europe, says Xiufang. I have on my desk a piece of bintangor-faced plywood that made the journey from Xuzhou Zhongyuan, a large plywood and board manufacturer based in Pizhou, north of Shanghai, to Britain. Since getting it, I have been looking closely at my kitchen units. The veneers do look very similar, I am afraid.

Greenpeace, in a study published in 2005, established that at least three major concessions in Papua New Guinea supplied plywood to major builders' merchants in Britain, via China. Since the Greenpeace report, several British companies have announced that they would end imports of plywood from China. But I found it still widely available. Britain was in 2006 by far the largest European market for Chinese plywood. The British Tropical Timber Federation advises its members against buying timber sourced from Papua New Guinea, because it will almost certainly have been logged illegally. But that is hard for a timber yard to comply with when the timber is simply labeled "Made in China." And harder still for you or I buying plywood or furniture or flooring.

Scott Poynton, head of the Tropical Forest Trust, says things can change if a "chain of custody" can be established for timber. He says the big DIY stores have the commercial muscle to demand Chinese timber mills find out and reveal where their timber comes from. Concerned customers, he said, should be looking for the label of the Forest Stewardship Council, which certifies sustainably logged timber. But even where that is not available they should be demanding that their timber suppliers can show the origins of their timber.

Making that happen in practice should become easier as agglomeration in the industry leads to the formation of larger timber combines in China. Timber combines with names and reputations to defend. Take plywood. Major Chinese companies in the plywood business include Happy International, which is based in Hong Kong and produces everything from plywood and veneer to parquet flooring and table tennis paddles. Another is Shanghai Xingaochao, owned by Tao Xinkang. Tao set up a furniture workshop in Shanghai in the late 1970s. Today he has twenty thousand employees and produces over 800 acre-feet of plywood a year. He has been listed among China's twenty richest entrepreneurs. Because of their size and visibility, such companies are easier to hold to account than small fly-by-night operations.

There are other important players in Scott's "chain of custody"— companies farther up the supply chain, nearer to where the deforestation is taking place. A lot of Happy International's logs, for instance, reach China via ships run by Pacific King Shipping of Singapore. A company prospectus that I found on the Internet says: "Our group's vessels are mainly engaged in the transportation of logs from the Solomon Islands, West Africa, Papua New Guinea and Southeast Asia to PRC [People's Republic of China] and India." It said the company shipped $19 million worth of logs in 2005.

It is not illegal to do this. Knowledge of illegal logging may be widespread, but guilt is hard to pin down. The paths taken by tropical timber, even in a simple journey across the South China Sea, can involve many layers of agents and a great deal of paperwork. But Poynton says he has special mistrust of the middlemen of Singapore and Sabah and other ports along the route from the forests to China. There is, he says, a small island off the coast of Sabah where the local Chinese Chamber of Commerce is notorious in the trade for issuing phony certificates of origin for millions of cubic meters of stolen wood from Indonesia, and possibly also Papua New Guinea. "The traders pay the chambers of commerce to issue the certificates as the ships go sailing by." In such an environment, establishing a traceable chain of custody is still going to be a hard task.

The Great Mall 16
Toothpicks to Placentas,
Everything Must Go

A young niece of mine has taken to declaring whenever she picks up a toy: "It seems to me that everything comes from China." Even eight-year-olds read the labels, it seems. And those toys, some of them Christmas presents from me, do all come from China. It's not just toys, of course. It's everything: big and small, smart and stupid, adult and juvenile. For the bottom-end stuff, I recommend a visit to Yiwu in southern China—about four hours' drive from Shanghai. Forget the Great Wall: this is the Great Mall, the bric-a-brac capital of the world. Or, as they put it there, "the largest wholesale market for small commodities on the planet."

I recognized things that have been cluttering up my house for years, things that turn up in every garage sale and Christmas stocking—and every charity shop just after Christmas. Toothpicks and tennis racquets; mothballs and candles that sing "Happy Birthday"; clogs and car jacks; walking sticks and rat traps; Russian dolls and fake African carvings. If you want to go to one place where the world's stuff comes from, go to Yiwu.

Yiwu is a town of about a million people. Not big by Chinese standards. They have a hundred that big. But this is where the world's retailers come to fill their stores. Where 80 percent of the world's Christmas decorations reach the global market, and 60 percent of children's toys. Christmas is made in China and sold in Yiwu. Every day in the run-up to Christmas, more than a thousand container loads of goods sold in its markets leave the city. The taste of much of it is frankly appalling, from the mantelpiece clocks with Crucifixion scenes behind the hands to the replica toy guns to aisle after aisle of children's clothing in migraine-inducing yellow and pink.

The city is dominated by a series of covered markets that have a combined area four times the size of New York's Grand Central Station and are occupied by thousands of tiny booths. Many appear to be selling the same things. But the accumulation is stunning. The city fathers claim there are fifty thousand booths selling more than 300,000 product lines. There are whole stores dedicated to selling different sorts of pacifiers and dusters and Halloween witches and purses and key rings and things to hang in car windows. There is an entire market devoted to socks, though I couldn't find a pair I liked.

And outside the covered markets, it is more of the same. God help you if you want to buy groceries. Instead, there are whole streets, whole neighborhoods, dedicated to selling individual items. If you want to buy a belt, you have a choice of a million, at least. Then there is bra street, zipper street, plastic products street, timber street, computer street, stationery street, necktie street, artificial flowers street, towel street, scarf street, calendar street, picture frame street, and Christmas crafts street. To name but a few.

Up near the prestige international market there is street after street of shops full of beads: plastic and wood and stone and glass, all different sizes, in bags and buckets and no doubt by the container load if you wish. There are single shops containing a million or more beads, and the entire neighborhood cannot contain fewer than a billion. Probably there was one for everyone on the planet, certainly one for every woman. It is the biggest bazaar in the world. By one estimate there are some eight thousand permanent foreign buyers in the city, constantly scouring the stalls for produce to ship home. There are so many Middle Eastern buyers that the city has its own Arab quarter. In the international market, there was a constantly changing display of the Commodity City Price Index. Helmets and ties were down, but locks and baby food were up. There seemed to be a run, too, on Father Christmases playing the saxophone. Yiwu has so many shops and booths that the main street is devoted not to department stores like in any other town, but to selling shelving systems and shopping carts.

Where does Yiwu get all this stuff from? Why does Santa do his shopping here? Because the province around it, Zhejiang, has made itself into the world capital for making … stuff. Suzhou may make the computers; Zhangjiagang may turn the world's rain forests into furniture, flooring, and plywood. But around here they seemingly make everything else. There are large factories, but also millions of peasant farmers and their families running backyard workshops. Most of these cut-rate capitalists have huddled together. Towns you have never heard of have captured great chunks of the world market in specialized and not so specialized products.

Take Qiaotou. It has two hundred factories making 60 percent of the world's buttons and zippers. That works out at 15 billion buttons and 200 million yards of zips—enough to stretch halfway to the moon. The president of the Great Wall Zipper Group proudly remarked recently that "there is almost nowhere else in the world that makes zippers." Without Qiaotou, the world's flies would be undone, and much else in disarray. Around 70 percent of the world's reusable cigarette lighters come from nearby Wenzhou. The Tiger brand is made here by some three hundred small workshops. Meanwhile another town, Cixi, makes a similar proportion of the world's disposable lighters, as well as millions of electric irons and air conditioners and washing machines, and billions of ball bearings.

No place in the world makes more drinking straws than Yiwu itself. The city also makes 3 billion pairs of socks a year, but is rather put in the shade by Datang, half an hour down the road, which makes 8 billion pairs. One in three of the world's socks comes from here. Datang is the base for the Zhejiang Stocking Company, started in the early 1990s by Dong Yong Hong, a former schoolteacher who began by selling at the roadside. She got big, but much of her output is subcontracted to legions of home workers in the town.

A billion or more people around the world are wearing shoes made in Wenling. One in every three men is reckoned to have a tie in a drawer somewhere from the 300 million made annually in Shengzhou. Hangji, in the eastern suburbs of Yangzhou, is the purveyor of 3 billion toothbrushes a year. Here are 1,600 backstreet producers and one giant, the Sanxiao Group, recently renamed Colgate-Sanxiao. Elsewhere in Yangzhou, there are 3,500 households engaged in producing the cheap disposable bits and pieces you find in hotel rooms, like slippers and folding combs and shampoo and shavers. Oh, and there is a toy factory with forty thousand employees. It has probably filled the toy cupboard of my niece. All this and more comes via Yiwu.

On the outskirts of Shenzhen, I stumbled on the world's largest market for fake oil paintings. At the Dafen Oil Painting Village, eight thousand painters work in eight hundred studios night and day to churn out imitations of the great masters—and some other surefire sellers. I have often wondered where those sorry paintings on the walls of hotel bedrooms come from; now I know. In one street there was a shop full of Mona Lisas; another with wall-to-wall Warhol Marilyn Monroes; another specializing in Canalettos; and a fourth with identical portraits of George W. Bush. Farther down the street, I met a real artist who couldn't understand why customers flocked to buy the fakes rather than his original watercolors. The trouble was he charged twenty times more.

It felt and looked like a large black prune. But you can't be too careful. "What is it?" I asked.

"It's a birth sac—a placenta," came the stallholder's reply.

"What animal?"

A brief smile. "Oh, a human. You take it for women's problems, and to make you more beautiful."

I hurriedly put it back in the market stall. I had heard of women frying their own placenta for a postdelivery breakfast. But other people's afterbirths? That sounded more like cannibalism than medicine. But China sells everything.

I was in the heart of Hehuachi, the traditional medicine market in Chengdu, capital of Sichuan province. A giant hangar the size of a couple of soccer fields was packed with stalls selling herbs and spices, potions, and animal parts of every description. Some of it would not have been out of place in a Western street market on a Saturday morning. But there was much that was more exotic. Immediately around me there were hedgehog pelts reputed to cure rheumatism, dogs' kidneys to promote sexual arousal, newts to treat stomachache, and dried snakeskins for dunking in wine as a tonic.

There were all manner of deer and antelope parts—feet and tails, horns and penises—jars of caterpillar fungus incongruously labeled in English, and boxes full of tiny crabs and dried seahorses, which, the stallholders insisted, would bring me "youthfulness." Other delicacies on sale nearby included bat feces, pangolin scales, seal genitals, the tiny fallopian tubes of frogs, toad venom, cuttlefish bones, and dried geckos and leeches arranged in rows like so many chocolate bars on a candy counter.

Hehuachi is one outpost of a fast-growing trade in traditional wild medicines that appears to have no limits. Based in China, it is spreading around the world. Many Westerners have tried to dismiss traditional Chinese medicine as a witch's cauldron of false remedies and bogus aphrodisiacs. But the truth is different, says Rob Parry-Jones, of Traffic, which monitors China's trade in endangered species on behalf of conservation groups such as the World Wildlife Fund (WWF).

Many potions have been found successful in epidemiological trials, he agrees. Research conducted for the WWF at the Chinese University of Hong Kong has found that both rhino and saiga antelope horn are potentially lifesaving cures for fevers and convulsions, for instance. (A flat contradiction of statements put out by the WWF over the years.) And Western drug companies have isolated a surprising number of active ingredients in Chinese medicines for use in their products. The Chinese treat gallstones with bear bile. Western firms have discovered that bear bile contains tauro-ursodeoxycholic acid, which in tests dissolves gallstones. (The only Chinese bear whose bile does not contain this acid is the giant panda, which is also the only bear from which the Chinese do not extract bile.)

Likewise, a 1,500-year-old Chinese treatment for malaria uses a daisy called artemisia, or wormwood. Its active ingredient is artemisin, which has recently been adopted by Western doctors. It works by reacting with the high iron concentrations in the malaria parasite, releasing free radicals that kill it. Another plant used in China to fight asthma contains ephedrine, a stimulant prescribed in the West for the same condition. And Western doctors recently patented a version of a Chinese remedy for eczema that uses the root of the peony shrub, apparently to strengthen the immune system.

The foremost chronicler of China's pharmacological history, Cai Jing Feng of the China Academy of Traditional Chinese Medicine in Beijing, is also a doctor trained in the Western tradition. He says that the philosophies behind the two traditions are very different. Western doctors generally treat the disease or the diseased organ, while the Chinese tradition is to treat the whole body to bolster its defenses. But whatever the philosophy, the treatments are often much the same.

Even the ancient Chinese medical ideas that Western scientists find most difficult to cope with, such as the notion of opposing forces of yin and yang within the body, turn out to have a physiological reality, said Cai. The yin and yang conditions diagnosed by Chinese doctors correspond to what Western doctors identify as disturbances in chemical messengers in the hormone system known as cAMP and cGMP. "When Chinese doctors describe yin and yang as being out of balance, Western doctors see a change in the ratio of the two chemical messengers in the body," says Cai. "In the yang condition, cAMP is low and cGMP is high. In the yin, it is the reverse."

In China, you are what you eat. And more and more Chinese are using traditional medicines as food. Whatever the potential risks from overdosing on active ingredients, the prevailing view is that the richer you get, the more "health food" you should eat.

The soaring worldwide demand for Chinese medicines is becoming a major threat to the survival of wild products, from obscure herbs to bears and tigers. Conservationists once hoped that the demand for Chinese medicines would decline because they would be seen as old wives' remedies. Now it seems the old wives' remedies do work. So groups like the WWF are changing tack, demanding that threatened herbs be cultivated rather than being gathered from the wild, while substitutes are found for animal products like tiger bones and bear bile.

It will be an uphill task. Economists have put the total value of the booming Chinese medicine market in wild products at between $6 billion and $20 billion annually, 85 percent based on plants, 13 percent on animals, and 2 percent on minerals. I never did find out the going rate for human placenta.

Part Five

Mines, Metals, and Power

My Beer Can   17
Giant Footprints in Bloke Heaven

I can't say I drink beer out of a can by choice. Give me a glass or a bottle any day. But like most of the world, I have gotten used to the sound of the pull-tab, the feel of aluminum against my lips, and the slightly metallic taste of whatever is inside, whether beer or coke or iced tea. Even so, I wonder how this seemingly innocent, if sensorily flawed, activity impacts on the planet.

Aluminum is the most abundant metal in the earth's crust, but it is only rarely found in a form that can be abstracted and purified —so rarely that Napoleon III once gave a banquet in which special guests got aluminum utensils, while the hoi polloi had to make do with gold. But modern extraction and processing methods have eased the shortages, and today it is the second most widely used metal, after iron but ahead of copper. Aluminum is handy stuff. It conducts heat and electricity, melts at a relatively low 1,220°F, and resists corrosion. It is also lighter, stronger, and softer than most other metals, so it can easily be drawn into wire, rolled into sheet, or cast into any shape you want. And it is pretty. For more than a century, the cool sheen of aluminum has been seen as modern, an ideal material for stylish design in everything from furniture to electronic goods to jumbo jets to, well, the humble can. Aluminum was the most distinctive metal of the twentieth century. But it also helped create another defining feature of the twentieth century—greenhouse gas emissions. For it takes a very great deal of energy to turn the ore into pure aluminum. The amount of electricity needed to make one beverage can will run a TV for three hours. And the world uses 250 billion cans a year. I went to Australia, one of the biggest producers, to explore the footprint of my beer can.

The story starts at Weipa on the Gulf of Carpentaria in northern

Queensland. This is the homeland of the Alngith people and, until the miners showed up in the 1960s, one of the last untouched wilderness areas in Australia. They have turned the wilderness into one of the largest single manmade sores on the planet's surface. The mining area established by Rio Tinto, the world's second-largest mining business, stretches for more than 60 miles and has engulfed three aboriginal communities. All told, the company has a license to strip-mine over 1,000 square miles of the bush and remove more than a billion tons of aluminum ore, called bauxite. That is enough to make around 250 cans for every person on the planet. Right now, a tenth of all the world's bauxite comes from here.

Digging up the ore is easy. It comprises a layer of red pebbles about 4 yards thick beneath about a foot and a half of soil. Huge digging machines scrape away close to 4 square miles of bush every year, much of it virgin woodland, to remove more than 26 million tons of pebbles. Rio Tinto puts back the soil afterwards, and vegetation will slowly recolonize the mines. But the new land is several meters lower than before.

After washing to remove soil and low-grade ore, some 17 million tons of bauxite a year goes to the Lorim Point Wharf, where it is loaded onto bulk carriers for a five-day ride around Cape York, Australia's northernmost point, and through the Great Barrier Reef to the port of Gladstone on Queensland's east coast. No vessels have yet hit the reef, which is a World Heritage Site. There would be an outcry if they did.

Gladstone wallows in another ecological mess, created as the town annexed once extensive mangrove swamps across a river delta. It has a nice harbor, where you can take boat trips out to the southern end of the reef. But it also has a cement works, a nickel refinery, a nitrate factory, and a mothballed operation for extracting oil from local shale. And it vies with Newcastle, down the coast in New South Wales, and Richards Bay in South Africa for the title of the world's largest coal port. Some 75 million tons a year come here by rail from mines in the Queensland interior. Burning that coal ultimately puts into the air about 220 million tons of carbon dioxide a year.

Most of Gladstone's coal goes to fuel steelworks in the booming economies of east Asia. But not all. For Gladstone also has Queens-land's largest coal-fired power station. And most of the power from the station is consumed by the masters of the town: Rio Tinto. For, where Queensland coal meets Weipa's bauxite, Rio Tinto has created in Gladstone the world's largest bauxite-processing operation.

Getting off the plane, I met Rick Humphries, head of climate change at Rio Tinto Aluminium. He is from the new school of big-industry environmental executives: cheery, anxious to proclaim that he once worked for Greenpeace, and at pains to be candid. "Aluminium processing is responsible for half of Rio Tinto's carbon foot-print round the world," he began. "And it is mostly because of operations in this town." We set off to check out the footprint, be-ginning at the waterside, where we watched a backhoe offloading red pebbles from the *RTS Pioneer*, a ten-year-old Japanese bulk car-rier. The 66,000-ton cargo was enough to keep production going for about two days. A conveyor took the pebbles to one of two re-fineries, where they are turned into aluminum oxide—alumina, as it is known in the trade. From there, the oxide is smelted into pure alu-minum ingots.

Rick calls Gladstone "bloke heaven." There is beer (in aluminum cans, of course) and fishing in the bay (from aluminum boats), and lots and lots of blue collar jobs. But at the refinery, after my safety in-duction, I met technical manager Anne Duncan. She smiled at my surprise. Female faces are rare in the town, rarer still inside the fac-tory gates. Anne hadn't planned on a career making blokes' beer cans. She trained as a nuclear engineer, but, she says, "I showed up for my first day's work on the day the Chernobyl reactor blew up, so I had a better idea and decided to retrain as a chemical engineer." Anne is the daughter of one of the leading academics in the aluminum business, John Duncan. My files contained a fading photocopy of a paper he wrote years ago on the life cycle of the aluminum can.

Anne drove me round the refinery. It is not the kind of place where you would just walk around: big and with an undergrowth of pipes and cables and vats and boilers as dense as any mangrove thicket. The bauxite pebbles from the conveyor are milled here into a powder of various aluminum-containing minerals. The powder is boiled up with caustic soda shipped in from the United States (around 1 ton of caustic for every 9 tons of bauxite), and the resulting liquid is drawn off into tanks where, after a few days, crystals of aluminum oxide form. To be turned into pure metal, the aluminum oxide goes to one of several Rio Tinto smelters scattered round the world, in New Zealand, Tasmania, Wales—or direct by conveyor across Gladstone to Australia's largest aluminum smelter, on Boyne Island. I took the short route.

Smelters are huge electrical furnaces, and this single plant uses as much energy as a city of a million people. The aluminum oxide is melted and dissolved into a conducting fluid, and then electricity is transmitted through the thick, hot mixture from giant carbon terminals, known as anodes. The electric current is a staggering 150,000 amps. It strips the oxygen from the aluminum and bonds it to carbon from the anode, making carbon dioxide and a sludge of pure aluminum.

The sheer scale of the operation took my breath away. It takes place in three cavernous smelting halls, each close to 3,000 feet long. Inside them are thirteen thousand anodes, each weighing more than a ton and the size of a child's coffin, arranged into five hundred cells that smelt hundreds of tons of molten aluminum at a time. Over the road is another factory almost as large, dedicated to making the carbon anodes from coal tar, pitch, and coke. To make them fit for conducting the vast amounts of energy, they are baked at more than 2,000°F for fourteen days.

The smelting process, known as the Hall-Héroult process after its nineteenth-century inventors, is largely enclosed. The operators would fry otherwise. But occasionally I could see the fiery glow as remote grabs lifted a spent anode out of the molten liquid. Humans were like tiny ants in here. We occasionally veered away from walls of heat coming from red-hot spent anodes standing to one side, cooling. Don't touch, said Alan Milne, the production manager. We hardly needed reminding.

The pure molten aluminum is eventually poured into molds. The cooled ingots head off around the world to be turned into all manner of goods. A lot of it goes to China. Gladstone aluminum may well be in my computer, my digital camera, my phone, and my printer. Gladstone also helps make European Airbuses. So I probably flew to Australia courtesy of this extraordinary inferno. The whole process had the feel of something primordial, or at least a throwback to Victorian engineering. Alan agreed. "This plant is relatively new, but the process hasn't really changed in a hundred and twenty years, since it was invented." Aluminum may look sleek and modern, but it is made in a way unchanged since the horse-drawn carriage.

Making aluminum has a huge environmental footprint. There is the wrecked wilderness of Weipa, of course. The two giant refineries converting bauxite to aluminum oxide create more than 5 million tons of waste slurry that is thick with both iron from the bauxite and caustic soda. Caustic soda is extremely alkaline. It burns. To neutralize the slurry partially, Rio Tinto mixes it with more than 18 million tons of seawater a year before pumping the resulting "red mud" into reservoirs several miles wide, which are scattered across the delta. They dominate the view as you fly in and out of Gladstone. "We still don't know what to do with them," admitted Rick. The iron could become a valuable resource one day. But nobody has yet found an economic way of extracting it. So there it sits.

Meanwhile, everyone hopes the reservoirs do not leak and contaminate the rest of the delta. That would be curtains for the remaining mangroves. The big fear is that one of the cyclones that regularly form in the ocean off Queensland will make landfall here. A full cyclone hasn't hit Gladstone for nearly a century. But on the wall in the smelter control room is a large map constantly updated to show passing storms.

And while we wait for the cyclone, Gladstone has a daily impact on the whole planet through its great, galumphing carbon footprint. Aluminum smelting requires more energy than any other metal process. Worldwide, the industry accounts for about 2 percent of electricity consumption. The Boyne smelter takes a constant supply of 900 megawatts of power, fizzing down four power lines from the Gladstone power station. The power station couldn't be worse. It is thirty years old, burns coal, and has a thermal efficiency of around 30 percent. That is pathetic by modern standards. It means that less than a third of the coal burned actually makes useful energy. As one joker put it, Gladstone power station is a giant machine for converting coal into carbon dioxide—with a little electricity as a byproduct.

The Boyne smelter also produces its own carbon dioxide as part of the production process. Adding both the power and production sources together, Boyne emits more than 18 tons of carbon dioxide for every ton of aluminum. In 2005, the smelter was responsible for more than 10 million tons of carbon dioxide emissions. Looked at another way, the smelting of aluminum to make every beer can here generates 260 grams of CO_2—enough gas to fill three hundred cans.

The electricity demands of aluminum smelting are so great that smelting companies will locate almost anywhere to get cheaper supplies. Rio Tinto plumps for Queensland coal at Gladstone. But its reliance on coal—which produces more carbon dioxide than other fossil fuel—is unusual and an increasing public-relations liability. Globally, more than half the power for aluminum smelting comes from hydroelectricity. This started when the United States built the Grand Coulee and Bonneville dams on the Columbia River during the Second World War to make the aluminum for sixty thousand fighter aircraft. They helped turn the tide on the Western Front. Ever since, many of the world's largest dams have been built to meet the smelters' bloated demands for power.

The construction of those dams has been the source of bitter disputes. To provide the power for Rio Tinto's smelter in Tasmania, the island's government filled Lake Pedder, one of the country's greatest natural treasures, to twenty-five times its natural size. UNESCO later described this as "the greatest ecological tragedy since the European settlement of Tasmania." The Ghanaian government flooded an area of its country the size of Lebanon to generate power for the Kaiser Corporation's smelter. Brazil inundated more than 900 square miles of Amazon jungle to create the Tucurui Dam, which powers smelters at Vila do Conde and São Luis. Canada's controversial dams on the Indian lands of Quebec power Alcan's smelters in the province. And it goes on. Alcoa, the world's largest aluminum producer, has just finished a controversial new dam in Iceland that captures meltwater from Europe's biggest glacier to power a smelter built largely with Polish labor on Iceland's east coast.

Clearly there is a serious downside to many hydroelectric dams. But as concern grows about global warming, fossil fuels look even worse. About 30 percent of the world's aluminum is still smelted using coal, and Rio Tinto probably does more of it than almost anyone. This made Rick's job as the company's climate czar a bit sticky. "We'd like to do more with renewables," he told me. But the reality, he admitted, is that Rio Tinto has no intention of giving up cheap fossil fuels. At Gladstone, it has a long-term contract to take power from the city's power plant. It half owns the plant and gets power for about half the rate paid by other industries.

But Rio Tinto is hedging its bets. It is planning to build a new smelter in Abu Dhabi that could handle a lot of Weipa's output. It will be powered by local natural gas. Sandeep Biswas, managing director of the company's aluminum operations, says, "The Middle East is fast becoming a key region in the global aluminum-smelting business." Why? It is a no-brainer, according to Rick. "They are outside the Kyoto Protocol." Australia originally joined the United States in reneging on the Protocol, so it failed to adopt targets for cutting its emissions of greenhouse gases such as carbon dioxide. But in 2007 a new government announced its intention to sign up. That is bound to have serious implications for big emitters like Rio Tinto. So the industry is preparing to make a speedy exit to countries that will be spared targets. Places like Abu Dhabi.

This is—how shall I put this?—not fully consistent with Rio Tinto's claims to be taking a greener outlook on life. Natural gas may be better than coal. But burning fossil fuels to make electricity to smelt aluminum is a rogue activity that should be shut down.

Should I boycott aluminum cans? Maybe. But wait. Could there be a path to redemption for aluminum? For one thing, aluminum is light. Everywhere that it replaces steel, for aircraft wings or beverage cans loaded on a truck, it saves energy. Of course, that gain has to be balanced against the emissions from making the aluminum in the first place. But one study found that the extra carbon dioxide generated by making aluminum to replace steel in cars is offset by the reduced emissions from the lighter vehicle within the first 15,000 miles—and after that it is downhill all the way.

In addition, aluminum is an ideal metal for recycling. Aluminum smelting uses so much energy that its producers have long since understood the virtues of "mining" waste metal for new manufacturing. Since the Hall-Héroult process was developed in the 1880s, it has produced more than 770 million tons of aluminum. But only 220 million tons of that has been discarded. The industry estimates that the rest is still in use somewhere. Metal originally smelted for the first bicycles or modernist furniture at the start of the twentieth century, or to make canteens for soldiers in the trenches of the First World War, is now in modern aircraft or computer casings or window frames—or my beverage can.

The industry plays this green card for all it's worth. Quite right too. Recycling aluminum takes only a twentieth as much energy as smelting new metal. So I set off to find out what happens when I chuck a can into a recycling bin.

Usually it ends up at Europe's only dedicated aluminum recycling plant, at Latchford Lock on the Manchester Ship Canal in northwest England. Novelis, the international company that owns the plant, reprocesses 35 billion cans a year around the world, almost a third of them at Latchford. Novelis is also the largest producer of aluminum sheet for beverage cans. It is closing the loop on cans.

My taxi ride to Latchford Lock ended in a front lot filled with bale after bale of squashed cans. I snooped around for a few minutes among the Stella and Murphy's, Coke and Sprite, Bulmers and Heineken, Foster's and Cobra, till I found the general manager, Mike Killen. He said the lot held ten days' supply—200 million cans altogether. In the boardroom we opened cans of Sprite. Latchford's aluminum recycling opened for business in 1942 as part of the war effort, Mike said. At the same time as the United States began big-time smelting at Grand Coulee, the plucky Brits were handing in their aluminum saucepans to be turned into fighter aircraft.

From the lot, cans go up a conveyor into the plant. I threw on my Sprite can. First it was broken into fingernail-size pieces by a giant hammer; next a magnet removed any steel; and then a blast of super-heated air stripped off the paint and lacquer. Then came the heart of the operation. The two furnaces each hold nearly 100 tons of aluminum, or almost 3 million shredded cans. Here the aluminum melts. A quick stir, and the molten metal is poured into molds, where a shower of water cools and solidifies it into new ingots. Each is 9 yards long, weighs over 28 tons and contains the molten remains of 1.6 million cans.

Looking at the ingots, I could have been in Gladstone. The same sleek slabs of pure metal. Better, said Mike. "We supply the cleanest can bodies in the world—even better than from primary ingots." Latchford keeps one furnace dedicated to cans, and the other to everything else. This preserves the distinct alloy preferred for cans, which contains small amounts of magnesium and manganese. Out at the back, a truck was ready to drive the next ingot, containing my old Sprite can, to the port of Goole on the English east coast. From there it will go by barge across the North Sea and up the Rhine to the world's largest aluminum rolling mill, outside Düsseldorf. The metal from my can might come back to supply UK can makers, and then reappear on my local supermarket shelf or in the pub fridge within six weeks. Equally, it could be in a Spanish bar or a Greek taverna or a Warsaw grocery store.

In 2006, Western Europe passed a milestone. For the first time, more than half of all cans sold came back for recycling. North Americans and Europeans normally pride themselves on being world leaders in recycling. But not with aluminum cans, which are sufficiently valuable that the poor urbanites of developing countries spend a lot of time finding and selling them on to recyclers. Globally, some 60 percent of aluminum cans get recycled. But in Western Europe and North America the figure is only 50 percent (U.S. recycling rates have actually gone down from over 60 percent in the 1990s). Within Europe there are sharp differences between countries. While Norwegians hit 93 percent, and Finns and Swiss 88 percent, the Brits manage a paltry 41 percent, making them among the world's worst can recyclers.

Besides being an ecological crime, this is also a failure of the market. Because cans are cash. Novelis pays a bit over one penny for a used can. Aluminum cans make up only 1 percent of the contents of our garbage cans (that's not counting the metal bin itself, of course), but they represent 25 percent of the potential cash-recycling value. For years, environmentalists have rightly seen aluminum smelting as a pariah industry. But if cans could be "mined" to the extent that the Scandinavians achieve, then its image could be transformed. If we recycled all our aluminum cans and foil and electronics casing and the rest, then we would scarcely need to mine the metal anymore. Aluminum could be the "greenest" metal in use, rather than the most polluting. As I left the Latchford plant, I added another can to the bales in the lot, and waved it goodbye.

Shock and Ore   18
Where My Metal Comes From

Remember the "limits to growth"? Almost forty years ago, apocalyptic environmentalists warned that we were about to run out of resources. Copper and tin and other key metals were going to disappear within a few decades. The Massachusetts Institute of Technology predicted that a scramble for the last metals and oil would prove so costly that "the industrial base will collapse." Civilizations would crash. Well, they haven't yet. The prediction of an exponential rise in demand for resources proved false. Since then, prices of key metals fell rather than rose, and a new optimism took hold. If metals ran low, we would always find more, or technology would create substitutes. The road to continued economic growth was clear.

But now the rapacious demands of surging economies like China are beginning to undermine the new optimism. At current rates of demand, there are thirty years of antimony and silver left, forty years of tin, and sixty years of copper. The date of peak oil production may be close. And, most worrying perhaps, a whole suite of new metals has become vital to the twenty-first century—many of them in short supply even as they are being exploited for the first time.

Our world now depends on indium, with a projected thirteen years of supply, on gallium and hafnium and terbium and ruthenium and, thanks to the mobile phone, tantalum. The -iums are in charge. Of course, technology may come to our rescue again, cutting our requirements, finding alternatives, or increasing minable reserves. But then again it may not. Or not always. The question is being asked again: are we running on empty?

For thousands of years, through the iron and bronze ages to the industrial revolution, metals have been the benchmarks of our economic progress. Finding metals has driven many of our conquests.

The Romans first came to Britain to extract tin. Gold and silver drew Europeans to the New World. Two-thirds of all European investment in Africa prior to the 1930s went into mining. And not too much has changed today. Mining is the world's fifth-biggest industry. The biggest mining corporation, the Australia-based BHP Billiton, formerly Broken Hill, has a turnover of $125 billion, roughly the same as Finland. Close behind are Rio Tinto and Anglo American, both headquartered in London.

The biggest rock-shifting businesses are iron, which removes more than a billion tons of ore a year, and copper, which removes close to 8 billion tons to make just less than 18 million tons of metal. Most copper now comes from a handful of vast mines. Utah's century-old Bingham Canyon is the largest manmade hole on Earth, 2.5 miles across and about a mile deep. During the Second World War, Bingham supplied a third of all the copper used by the Allies. But the world's biggest source of copper today is Chile. The country has two-fifths of known world reserves (enough to keep us going for twenty years), and 8 percent of global supplies currently come from the Escondida mine in the Atacama Desert in northern Chile, which removes some 385 million tons of ore a year.

Mining has a big and brutal footprint on the land. Over the past century, the industry has displaced 100 million people, and laid waste innumerable precious habitats. The extraction and refining of ores often requires toxic substances, like cyanide and mercury, that pollute land and rivers. A ton of mercury is released into the Amazon for every ton of gold extracted. Worldwide, metals smelting pollutes the air with more than 100 million tons of sulfur dioxide, the main cause of acid rain.

And, while most human economic activities reduce their footprint with time, the mining footprint just keeps on growing as the available ores become less concentrated. At the start of the twentieth century, copper mines like Bingham Canyon were extracting ore with a metal concentration of 3 percent or more; by the end of the century, that figure was down to 0.6 percent. Tungsten is now taken from 0.25 percent ores, zinc from 0.05 percent, and gold from 0.02 percent or less. The holes in the ground, not to mention the spoil heaps and tailings reservoirs, grow ever larger. The ecological and social havoc increases.

Some talk about the ore taken from the earth as the materials "rucksack" of everyday products. By that they mean that every item we own made from stuff hewn from the earth carries with it the burden of all the waste ore that had to be mined to produce it. This book started with gold. My wedding ring took more than 2 tons of ore to produce. So its rucksack weighs more than 2 tons. Modern mobile phones weigh about 2.5 ounces, but their materials rucksack is 165 pounds. My watch clunks along with 44 pounds, heavier than most grandfather clocks of old. And my PC computer has a 1.5-ton rucksack. That's a lot of computer. To make a ton of aluminum typically requires 4–6 tons of bauxite ore. A ton of iron requires the mining of 15 tons, and a ton of copper a staggering 420 tons. My personal annual materials rucksack, as a typical Westerner, weighs 55 tons.

How does this 55 tons break down? On average, 16 tons comes from extracting fossil fuels like coal. Another 13 tons is accounted for by metals—of which copper at 2 tons, and iron, tin, and gold at a ton each, are the biggest components. Then come construction materials such as stones, sand, gravel, and limestone, which make up a hefty 10 tons. After that biomass, including our food and other crops like cotton and timber, takes 7 tons. Behind those come losses of soil to erosion, which is estimated at 4 tons, excavation and dredging at 3 tons, and nonmetallic industrial minerals (everything from diamonds to phosphate rock for fertilizer) at 2 tons.

This analysis, from the Wuppertal Institute in Germany, does not include water, which I return to later. Suffice to say, my rucksack should also, strictly speaking, include most of the contents of an Olympic-size swimming pool.

What do we use these laboriously extracted materials for? Most iron makes steel, and most steel is used to create the structures within which we live and work, and travel—homes, offices, factories, and cars. There is more steel than anything else in cars and planes, though aluminum is catching up. Almost half the world's copper goes into electric cables, plumbing, and heating systems in buildings. About a third goes into phone systems. Then there is lead, two-thirds of which goes into car batteries. Zinc, chromium, and nickel are all widely used in alloys to make corrosion-resistant, stainless, or just plain shiny steel. Tin goes into cans, solders, and alloys like bronze.

Some materials are only found in economically viable quantities in a few places. South Africa has 88 percent of the world's platinum, 40 percent of the world's extractable gold, and 35 percent of its chromium. Most of the rest of the world's chromium is in Kazakhstan—a fact that inexplicably escaped Borat's attention. China has 60 percent of our antimony, which is widely used in the ubiquitous electrical conducting devices known as semiconductors, and in flame retardants. There may be only twenty years' supply of antimony left. China also has 30 percent of our tin and 20 percent of our zinc.

All this makes resource politics interesting. We know all about oil politics. And the world of blood diamonds has become notorious. But what about phosphate politics? Phosphates are an essential nutrient in soils. Plants need phosphate to grow as much as they need water. It takes a ton of phosphate to produce every 130 tons of grain. There are no substitutes. If natural phosphate in the soils is in short supply, farmers must add phosphate rock. To that end, the world mines about 155 million tons of the stuff a year, mostly from the United States and China (which both consume most of what they produce) and Morocco. Or rather, and here is the rub, Morocco and the neighboring desert state of Western Sahara, which Morocco annexed in the 1970s.

Morocco and its neighbor have more than 60 percent of all the world's known economic reserves of phosphate—reserves that may run out by the end of the century. Morocco's hold over this critical resource may explain the world's reluctance to demand independence for the Western Sahara. Meanwhile the Moroccan royal family, through a process somewhat euphemistically termed privatization, has retained personal control of most of the country's mining operations for phosphate through their firm Omnium Nord Africain. Right now, much of the world continues to grow its food thanks to the favor of one North African royal family.

Prices of most mainstream metals have surged through this decade, as China in particular has sucked in supplies. Many old mines have been reopened. From Congo and the copper belts of Zambia to Indonesia and Peru, peasant miners have given up working on farms (where world prices have been in freefall) and switched to scrabbling through the rubble and tailings of old mines to find low-grade ore to process in makeshift smelters and refineries.

Meanwhile, the world is growing increasingly dependent on a range of metals that most of us never heard about in our school chemistry lessons. This is not what we were promised in the days (oh, just a couple of years ago) when people said that the information age would reduce our dependence on physical resources like metals. But we cannot escape that easily. Our dependence hasn't gone away, it has just changed. We have new metals fixes.

For instance, thirty years ago we decided to clean up the smog-producing chemicals in car exhausts by fitting them with catalytic converters. They filter out the pollutants using two metals: platinum and palladium. World demand for both has soared as a result, and catalytic converters take almost half current supplies. South Africa is the world's dominant source of platinum and, since it has most of the world's reserves, is likely to remain so. Meanwhile, more than half of the world's palladium comes from Siberian mines. The metal is mostly refined at notoriously polluting metals smelters at Norilsk on the edge of the Arctic Circle. These smelters are the biggest concentrated source of sulfur dioxide on the planet. For hundreds of miles round Norilsk, the trees are dead because of the acid they generate. In effect, we are destroying huge areas of Arctic tundra with acid rain so that the rest of the world can keep its city air clean. I can't say that makes me breathe easy.

Modern electronics, meanwhile, is adding to the list of obscure metals on which we depend. Ever heard of the ruthenium rush, the bismuth bonanza, or the indium stampede? Nor had I. But my office is full of these metals, and metals prospectors are scouring the world looking for more. Indium, for instance, is a soft metal that can be made into a very thin coating. It is built into a billion consumer devices a year, mostly the LCD displays of TVs and mobile phones.

From early 2003 to 2006, indium prices increased eightfold to $800 for a kilogram (just over 2 pounds) because of fears that economic sources (indium is largely a byproduct of zinc mining in China) might run out within a decade or so. Indium is also an ingredient in a material called indium gallium arsenide that is vital to a new generation of photovoltaic solar cells. With gallium also in short supply, the boom in the new solar cells could be short-lived.

Meanwhile, demand for bismuth (currently mostly produced in China) is surging because of its use as a safer substitute for lead in the billions of tiny drops of solder applied to consumer electronics goods. Its price has doubled, while that of ruthenium, used in resistors and disc drives, rose sevenfold during 2006. Ruthenium is found in the Ural Mountains of Russia and can be extracted from spent nuclear fuels. Then there is hafnium. It is in big demand as a substitute for silicon compounds in computer chips, and could be gone by 2017. And terbium, one of the rarest of the "rare earths," but a raw material for the boom in low-energy fluorescent light bulbs because it glows with a green light. The biggest reserves are in clay in southern China. They could be gone by 2012.

In an age of scarcity, hotspots of valuable materials become so hot that, particularly in poor countries, nature's bounty becomes a curse. It seems to stymie economic development, cripple governments, fill the pockets of rebels, and all too often pay for protracted civil wars.

Take Angola. Its rich resources make it potentially one of the wealthiest countries in Africa. But it has become synonymous with squandered wealth. After independence in 1975, it descended into civil war. Initially the United States, Russia, and South Africa armed the combatants. But when the Cold War ended and the superpowers lost interest, the local combatants kept going by plundering natural resources. The main rebel group, led by the late Jonas Savimbi, invaded northern Angola to grab the diamonds littering the riverbeds. He raised several billion dollars to buy arms. Meanwhile, the government in Luanda armed itself by selling off the country's huge offshore oilfields to big oil companies.

In Afghanistan, the Taliban and the Northern Alliance have kept operating over the past decade, whether in government or not, by selling emeralds, timber, and opium. Afghan opium poppies currently have a 90 percent world market share. In Colombia, rebels have prospered from cocaine sales and used the threat of sabotage to extort hundreds of millions of dollars from oil companies. And in Cambodia, the friendless Khmer Rouge kept going through the early 1990s by selling rubies, sapphires, and logging rights in their jungle terrain.

Once, we feared conflicts when resources ran out. But in truth conflicts are more likely where there are abundant natural resources. Michael Renner of the Worldwatch Institute, a Washington, D.C. think tank, calculates that a quarter of all conflicts around the world are either fought over, or largely funded by, the looting of lucrative natural resources. Today's rebels are often more interested in liberating diamonds than in freeing repressed peoples. Worse, they finance their start-up costs by selling the spoils of war in advance.

"Booty futures" have become the currency of conflict, as wannabe warlords sell resource rights to foreign corporations, mercenaries, and governments. In Sierra Leone in the 1990s, rebels got off the ground by selling diamond futures to Charles Taylor, a warlord and subsequent president in neighboring Liberia. When the rebels got within 20 miles of the capital, Freetown, the government hired mercenaries by offering them the same diamond fields.

Since the end of the Cold War, rebel groups can no longer expect funding from superpowers jockeying for strategic advantage. And abstract ideologies have become less potent. Moreover, particularly in the poorest states, globalization and the rise of the free market are reducing the power of governments. They control far less of their national economies. At the same time, rebels find it easier to sell their booty abroad. As a result, even successful rebel groups may conclude that there is little to be gained by seizing the reins of government when the rewards are few and when perpetual conflict maximizes the potential for pillage, extortion, and exploitation of labor.

David Keen of the London School of Economics argued in an influential essay, "Economic Functions of Violence in Civil Wars," that "we tend to regard conflict as a breakdown in a particular system." In fact, he said, it represents "the emergence of another, alternative system, of profit and power." If he is right, then the crucial resources we need to run our lives will increasingly fall into the hands of criminals and warlords. And the future for an ordered, stable world looks bleak indeed.

What should we do? How can we maintain our supplies of diminishing resources and reduce the minerals curse? One answer is recycling, turning waste products into the new mines. We cannot anymore chuck scarce metals in the landfill. If we do, we will soon have to mine the landfills. Or even street dust. That is a serious proposal, incidentally. Platinum can reach concentrations above one part per million in street dust, which makes it a richer resource than currently commercial copper ores. And palladium fallout in the soils around Norilsk is so heavy that it too may soon be worth extracting.

The world has always recycled some metals. Gold and silver, for instance. In the Middle Ages, church bells were melted down for bronze cannons in wartime—and afterwards, to celebrate victory, the cannons were turned back into church bells. Today, the economic case for recycling is becoming ever stronger. The world is not short of copper, but it makes more sense to recycle used metal than to mine the remaining ores. About 30 percent of the world's copper consumption is currently met by recycling. Likewise, as we have seen, most aluminum gets recycled. Again, there is no urgent shortage of bauxite, but recycling it makes sense, given the very large energy savings. Other metals with significant recycling rates include nickel (35 percent), chromium (25 percent), lead (72 percent), zinc (26 percent), and tin (26 percent).

Iron too—or rather the usual end product, steel—has also long been recycled. Between 70 and 90 percent of all the steel waste in North America and Europe goes for recycling. Modern electric arc furnaces take only a third as much energy to produce new steel from recycled steel. As the environmentalist Lester Brown puts it, "In the new economy, electric arc steel mills will largely replace iron mines."

My water footprint is huge, and so is yours. I have written a previous book about our use and abuse of water, so this is just a summary. But I still find the stats extraordinary. I drink only about half a gallon of water in a day, most of it in tea and coffee. We don't have a water meter in our house, but I reckon that, as a fairly average British water user, even after cooking and washing and flushing the toilet, I don't get through more than 40 gallons a day. But that is just the start of my water footprint. The scary numbers come when I look at the water it takes to grow my food.

To grow enough wheat to make a slice of toast takes 40 gallons. So my daily consumption is doubled, just to make some toast. A single portion of rice in my curry needs about 25 gallons; a pint of beer to wash it down takes more than 65 gallons. But for meat eaters and even dairy-product lovers, the news gets worse. Growing grain to feed enough cows to deliver a single quart of milk requires over 1,000 gallons of water, and it takes nearly 3,000 gallons to generate a quarter-pound hamburger.

For food grown in Britain, much of this water comes from rainfall, which at least makes it cheap. But increasingly Britons are buying food from distant lands, where rains are scarce, rivers are being emptied, and underground reserves pumped dry. So my water footprint becomes a global concern. The corn in my breakfast cereal probably comes from the American Midwest, where the underground water reserves are drying up fast. Increasing amounts of my meat come from cattle raised in sheds where they are fed with irrigated alfalfa. Even homegrown chickens are fed soy from Latin America.

If my sugar is grown by cane farmers in India, then, according to a UN report, those farmers often "use huge quantities of water in places where there is little rainfall...the poor farmers are left to fend for themselves." In Israel, 70 percent of the country's water is used to irrigate export crops like tomatoes that turn up in my supermarket; meanwhile Palestinians on the West Bank are told they cannot sink new wells to meet basic household needs because there is a water shortage.

And other farm products like cotton or leather take a huge toll. For my pair of leather shoes, it takes more than 2,000 gallons of water to raise the animal and turn its skin to leather. The cotton in my shirt takes around thirty bathtubs of water to grow. And if that cotton comes, as it may well do, from Uzbekistan, then it will be helping empty the Aral Sea.

Water is a renewable resource, of course. Ultimately nature will recycle it, through evaporation and rainfall. We are still drinking the same water that the dinosaurs bathed in. But today humanity is extracting water from the natural cycle on such a scale that it is often not available when we need it and where we need it. If a farmer has no water, then it is little compensation to know that it is raining somewhere else.

Two-thirds of all the water that humans take from nature is used for agriculture. By some estimates, 10 percent of that agricultural use of water—more than 810 million acre-feet of water, or twenty Nile Rivers—ends up in international trade. The water is not traded, of course; but the products of its use are. Economists sometimes call this "virtual water." And as countries across the world run short of water, this trade in the virtual stuff is increasingly important. The phrase is a bit of an abstraction, I agree. But it does show how water is becoming an increasingly vital resource, and those who have it have something of increasing value.

The biggest virtual water trades are in beef, soy, wheat, cocoa, coffee, rice, and cotton. All of them are global commodities that are often grown in poor, dry countries to meet the demands of people like me in richer—and often wetter—countries. So whenever I eat Thai rice, or burgers made from Costa Rican beef, or wear clothes made of Uzbek cotton, or sweeten my coffee with South African sugar, I may be emptying rivers or denying some farmer the water he needs to feed his family. Among the world's top ten exporters of virtual water are drought-stressed countries like China and India.

My own total water use is around 2,000 tons a year, or 100 times my own weight every day. If everybody in the world had the same requirements as me, then that would work out at 10.5 billion acre-feet of water, which is only slightly less than the total amount of accessible water flowing down the world's rivers in a year. No wonder they are starting to run dry. The natural water cycle may renew our water, but it does not do it fast enough to meet our demands.

I find it a shocking thought that the world is running up against real global limits on the supply of nature's most basic resource: fresh water. I could, in a pinch, live without indium and copper and tungsten and aluminum and iron and the rest. I could not live for one day without water.

Footprints in the Snow   19
Finding the Last Oil

George Tagarook, the Stratocaster-playing Inuit heartthrob of Kaktovik, was proud of his three gleaming fire engines. He showed them off in the driveway of his fire station in the small town on the north coast of Alaska, where he was both fire chief and vice mayor. Across the road were a new school, clinic, post office, and power plant. "All this," he said proudly, "is built with oil money."

Kaktovik lies within Alaska's huge Arctic National Wildlife Refuge. To the south is pristine tundra. To the north are the great ice floes of the Arctic Ocean. But a half-hour flight to the west lies North America's largest oilfield, operated by BP around Prudhoe Bay. For three decades, more than a thousand wells here have been pumping up to 2 million barrels of oil a day from beneath the tundra, and sending it down a 500-mile pipeline to Valdez on the southern shore of Alaska. From there, tankers take it to refineries around the world.

Prudhoe Bay is one of a series of remote regions of the planet where the fossil fuel industry has implanted its giant footprint. One of the places that keeps the world's gasoline tanks full and homes heated. One of the places that keeps me traveling around the world. One of the places responsible for heating the planet's atmosphere. And its extraordinary wealth is the reason why George can have three fire engines in a town of two thousand people where they have only four fires in an average year. Interviewing George was part of my inquiry into my oil footprint in remote parts of the world that few of us get to visit.

But George was worried. Oil production at Prudhoe Bay was falling, he said, as wells were exhausted. Revenues had been slipping, and he needed new fire engines. "Without oil, this place will go down," George told me. And in August 2006, after my visit, production was shut down altogether for a while, following the discovery of corrosion in the pipes and oil leaks onto the tundra around Prudhoe Bay.

Like most people in Kaktovik, George wants the U.S. government to lift its ban on oil prospecting within the wildlife refuge and allow the oilmen to move east to keep the oil and money flowing up here on the Arctic shore. With talk that the world is close to the start of a long-term decline in oil production, the case for opening up those reserves is being heard loud and often. But there is also a huge international campaign to prevent a further giant field being opened up on this pristine tundra, where the caribou and polar bears are already threatened by climate change. And in that battle, Kaktovik is on the front line.

The lives of the Inuit at Kaktovik and in other towns along the north shore have been transformed by oil. But not entirely destroyed by it. The second-biggest industry on the coast, after oil, is traditional whaling. There are eight old whaling families in Kaktovik. Each September, they set off in search of the three whales that the International Whaling Commission allows them to catch. The harvest is cut up and stored in underground ice cellars, and shared among the community—Stratocaster players and all.

On a bright Sunday morning, at Kaktovik's tiny Pentecostal Church, the Reverend Isaac Akootchook announced to the congregation, many of whom work for the oil companies, the date for the next distribution of whale meat and muktuk (a prized cut of skin and surface blubber). In his younger days, he said afterwards, he had been a whaler. He caught the largest whale ever landed at the village harbor—a 55-foot beast.

Isaac said whaling was once the heart of the community. But things started to change in the 1950s, when the U.S. military established an airfield and an early-warning radar post to spot Russian bombers flying in over the Arctic Ocean. He was among those thrown off the original town site to make way for the airstrip. The town still contained disturbing hints of how the military had treated its hosts. One sign in the community hall read: "If you remember having been subject to human radiation experiments, please contact..." Another asked for the help of polar bear hunters in providing bear organs for a study of chemical contamination.

But that legacy pales beside Kaktovik's current Faustian pact with big oil. Oil revenues course through the local economy. Average incomes here are far above those in non-oil towns in the state. And some years ago, the town invested some of its oil revenues in buying control of almost 250,000 acres of Arctic refuge that are believed to be rich in oil. The Inuit want permission to start pumping.

That's the official line, anyway. But I found a lot of whispered opposition. "This place is very divided about oil," said Carla Kayotuk, who runs the town store. Many fear what oil is doing to the environment. Isaac is against opening up the reserve because any expansion of the industry will damage whaling. Others fear the effects on their society. Alcoholism is one problem. Alcohol is banned by the town, but people say travelers recoup their airfares by smuggling in bootleg booze. The community occasionally evicts unruly people: four in the last three years. The exiles end up down the coast at Nuiqsut, described to me as an Eskimo "sink settlement."

Outsiders who have been long-term residents, especially those skeptical about oil's benefits, say they feel frozen out of the town's affairs. They are rejected as not being Inuit. And yet many of the town's leading families are far from purebred. The oldest citizen, eighty-nine when I visited, was Nora Agiak. She remembered traveling across the Arctic ice on dog-drawn sleds. She still wore moose-skin moccasins and a jacket of wolverine fur. She claimed to speak no English, yet neighbors said she was the daughter of a Scottish whaler. Kaktovik is not a town at ease with itself.

The Inuit are not the only people up here with an ancestral claim to the tundra. I flew to see a community 125 miles inland, on the southern edge of the wildlife refuge, that takes a very different view of big oil. Arctic Village has an airstrip, but no road, no hotel, no domestic plumbing, and certainly no fire engines. The people here are Gwich'in, North American Indians, and their lives remain much more communal. But the politics of oil is bringing the outside world in.

The timetable on the communal bulletin board listed the visitors that week: two environment groups, some press, a team trying to sell the village solar panels, and a couple from the Isle of Man on a world tour. Next up were some congressmen. But the first item on the board, in preparation for the visitors, read "village clean-up"—no doubt the reason for the huge pile of beverage cans behind the house of Sarah James, head of the Gwich'in intervillage committee, which was set up to fight the oil companies.

Arctic Village is one of a dozen communities of Gwich'in strung out along the migration route of caribou known as the Porcupine herd, after the name of one of the rivers they cross. The traditional ways of the Gwich'in are built around an annual hunt of the 110,000-strong herd as it moves from the snowy fastnesses of northern Canada, through the mountains near Arctic Village, and on to its summer calving grounds on the coastal strip of northern Alaska near Kaktovik, where the cottongrass grows richest and the ocean breezes keep the vicious local mosquitoes at bay.

For thousands of years the caribou have survived grizzlies, golden eagles, wolves, and human hunters. The oil industry has already eaten into grazing grounds around Prudhoe Bay. So far, the pastures of the Porcupine herd to the east are untouched. But the seven thousand Gwich'in people believe that if the oil companies expand into the wildlife refuge, the herd will disappear—and their own way of life too will be lost. "We are caribou people," said Sarah. "That is how we identify ourselves. That is how we feed ourselves."

Though different in many ways from the Inuit of Kaktovik, this community too is fractured. "The current generation is lost between two worlds," said village schoolteacher Mary Groat, who has since left Arctic Village. There is the world of the elders, underpinned by the Gwich'in language and the caribou hunts, and the world they see on TV or when they go to college in Fairbanks.

"Caribou hunting is now the main tradition we have left," Mary said. Many people send dried caribou meat to their exiled relatives in the cities. Caribou meat was Sarah's parting gift to our party. It is their nutrition, the wellspring of their communal culture, and their tie to the landscape. A young Gwich'in village official, Roland Tripp, sat in front of one of the council's computers. "We need cash, of course we do. But our caribou are more important to us than money. And if the oil companies come to the refuge, we will lose the caribou." Sarah puts things more simply. "We need the caribou herd like the Amazon Indians need the rain forest. It is our lives and culture." But that is to romanticize. The truth is more complex. At the communal barbecue, caribou meat and fish were supplemented by hamburgers, pasta, and a never-ending supply of sugary biscuits.

It was late June. My visit had been timed so I could see the Porcupine herd on its pastures. But the timing had gone awry. The herd was a month late. Heavy snows in the mountains—caused, paradoxically, by warmer winter weather—had delayed their migration. We camped on the edge of the coastal plain and saw the first males come by. The females were straggling behind, often dragging newborn calves across rivers swollen by melting snow. This had never happened before, Fran Mauer of the U.S. Fish and Wildlife Service in Fairbanks said later.

There were many tears in Alaskan schools that month. The children had "adopted" caribou cows fitted with radio collars and were receiving regular updates on their progress toward the pastures. But a quarter of the adopted animals died, along with many of their calves. Among them was Gus-Gus, a calf being followed by the children at the Arctic Village school.

Most biologists I spoke to in Alaska were convinced that the Porcupine herd was already under severe pressure from manmade climate change, caused in part by burning the oil from Prudhoe Bay. But now the animals face the ultimate threat—the appearance of oil rigs on their grazing grounds. Can they survive such an invasion? Nobody is sure. The greatest caribou herd on the planet, responsible for one of the last great mammal migrations, could be doomed, and with it the lives of their hunters. It was time to go and see big oil.

Flying west from Kaktovik along the Arctic north shore in a four-seater bush plane, bumping through a rainstorm late in the evening, I first saw the pipelines, then roads, and finally the lights and flares of Prudhoe Bay itself. This is a major industrial enterprise in the middle of the Arctic. Hundreds of millions of dollars are invested here. After landing at Deadhorse airstrip, I met BP's resident biologist, Ray Jakubczak. He said the local caribou don't mind the oil developments. He stopped the car. We watched a caribou dodging the trucks going to the wellheads, before ambling under the main pipeline linking the field to the port of Valdez in southern Alaska. It didn't seem too upset. "If we get permissions to develop in the refuge, it certainly won't be a wilderness anymore," Ray conceded. "But the wildlife will be fine."

Later, when I met caribou biologist Ray Cameron at the University of Fairbanks, he choked in disagreement. "Yes, you saw caribou round the oilfields. But they are not from the Porcupine herd, and you only saw males. The females and their calves keep well away from roads and oil developments. Around Prudhoe Bay they can do that because the coastal plain is wide. But farther east, which is where the Porcupine herd have been calving for thousands of years, and where BP wants to drill, the plain is much narrower. There is less room for them to hide."

BP says any future developments would have a much smaller "footprint" than Prudhoe Bay. There would not even be road links. Everything would go in and out by helicopter. But even so, the north slope is a precious place. BP's recent record of oil spills up here does not command confidence in its ability to keep its footprint small, whatever the good intentions. And nobody knows how the Porcupine herd would respond to its presence. It would be shameful for the herd to be destroyed for a few months' supply of oil to the world.

Whatever its failures at Prudhoe Bay, BP's footprint in Alaska is dainty compared to the way the bogs and forests of West Siberia have been trampled by Russia's quest for oil and gas. I have traveled to this region twice: once in the 1990s and again in late 2005. What was once the world's largest swamp—frozen solid in winter but with thick, bouncy peat bogs in summer—has been brutally scarred by big oil, Russian style.

Once, we could blame that on Communism. In 1926, in one of the most notorious statements of Soviet intentions toward nature, Stalin's favorite writer, Vladimir Zazubrin, declared: "Let the fragile beast of Siberia be dressed in the cement armor of cities, armed with the stone muzzles of factory chimneys, and girded with the iron belts of railroads. Let the taiga be burned and felled; let the steppes be trampled. Only in cement and iron can the fraternal union of all peoples, the iron brotherhood of man, be forged." And so it was.

Notions of an iron brotherhood may have passed into history. But not a lot else has changed. Now, Russia's new generation of capitalist oligarchs is continuing to trample the taiga. And, increasingly, it is Western consumers who are the market for the oil and gas coming out of here. The West Siberian oilfields now rank close behind Saudi Arabia. And the region is second only to the Middle East in its reserves of natural gas, which come down a vast network of pipes to keep much of Western Europe warm in winter. West Siberia's hydrocarbon reserves make up about a fifth of what the planet has left.

On my first visit, I landed at a Siberian city called Noyabr'sk, one of the major centers of the Siberian oil industry. Only incorporated in 1982, Noyabr'sk has a population of 100,000. Most of them are Ukrainians shipped in during the latter days of the Soviet empire. I visited with Western oil engineers brought in during a fit of glasnost to advise on the future development of the oilfields. During many hours of helicopter flights, we saw a landscape where the main features till three decades ago were reindeer tracks. But now it was covered with pylons, oil pipelines, drill rigs, a network of endless roads laid through forests or on wide embankments across lakes and bogs, and straight narrow gashes that marked the routes taken by old seismic surveys. No attempt had been made to run these different pieces of infrastructure along common corridors. Instead they snaked independently across the land, dividing nature into millions of fragments, each only a few hundred yards across at most. It was as Zazubrin decreed.

Pollution was everywhere. We flew through clouds of black smoke from gas flares and burning pools of waste oil. Our experts estimated that a tenth of all the oil pumped to the surface was lost before it left the oilfields. Our expedition visited one drilling pad unannounced. The pumps were pumping, but there was nobody there. The rig was surrounded by blackened forest where waste oil had poured into a creek and plastered tree foliage. We could see sheens of oil glistening on the lakes and bogs all around.

Nowhere on Earth, the experts concluded, has such a large hydrocarbon resource been exploited so quickly and so wastefully. The region round Noyabr'sk had twelve thousand oil wells—ten times more than the Prudhoe Bay oilfield. They had been drilled from several thousand large concrete drilling pads. Often the pads were less than 1,600 feet apart, each requiring its own roads, pipe, power lines, and waste dumps for the millions of tons of polluted water generated during drilling. Modern methods could replace twenty or more well pads with a single drilling point, the Western oilmen said. But their Siberian counterparts were "like cowboys, taking the cheapest and easiest oil in the cheapest and easiest way."

The huge oilfields around Noyabr'sk were then owned by a state company, Noyabr'skneftegas. They supplied one of the world's largest oil refineries, in the Siberian city of Omsk to the south. In 1999, both the oilfields and the refinery came under the control of Roman Abramovitch, the soccer-loving Russian tycoon who now lives in Britain and owns Chelsea Football Club. The reserves of the new company, Sibneft, were believed to be greater than those of Exxon, and the refinery's "light sweet crude" is sold to the West for jet fuel and low-sulfur diesel for cars and trucks.

Often, when I fly out of Heathrow or Gatwick or change planes at Schiphol or Frankfurt, my plane will be fueled with Sibneft's oil from the bogs around Noyabr'sk. In 2005, Abramovitch sold Sibneft at a profit of several billion dollars to the huge state energy conglomerate, Gazprom. So whenever I go and watch Chelsea's soccer players take the field, I reckon I am watching the team that Noyabr'sk's oil built.

Our trip to Western Siberia was extensive and, so far as I am aware, unique for independent outsiders. We flew northeast from Noyabr'sk, dropping in on an abandoned Stalin labor camp called Gulag 501 Construction Enterprise. Half a century ago, it housed political prisoners brought to work on an abortive railway project along the Arctic Circle known as the Great Stalin Railway. Now it held only ghosts. We continued to a tiny town called Krasnoselkup on the banks of the Taz River. Here they hunted sable and squirrel, fished the rivers, and logged the forests. I was offered the pelt of a big brown bear for $1,000. They didn't believe me when I said I could never take it home, because Customs would impound it and arrest me.

Russian engineers had just started drilling outside Krasnoselkup. The main test rig was about 100 yards from Devil's Lake, a major center for wildlife. The deputy mayor, Alexander Laetov, told me how the exploratory drilling was frightening away the reindeer and poisoning fish. (We weren't helping. Our ex-military pilot spotted a bear and chased the terrified animal through the wet landscape. It was like a scene from *Apocalypse Now*.)

We met a small tribe of native Selkup people camping on the banks of the Taz. There are only four thousand Selkup left in the whole of Siberia. They told the story of their centuries-long march north in the face of advancing Russians. In the woods they showed us their shaman's sledge-load of sacred objects. Among them were arrowheads and metal ornaments in the shape of birds and snakes, made, they said, in the days when the Selkup were a great people and smelted iron. But they and their reindeer were on the retreat once again as the oilmen moved in.

Since that visit, Gazprom has moved into Krasnoselkup in force. The area around Devil's Lake is now called the Yuzhno-Russkoye oil and gas field. Roads and pipelines and the rest of the infrastructure needed for a major gas field are being constructed. Production begins in 2008. The gas will go down a pipeline across Russia and beneath the Baltic Sea to northern Germany.

But the municipal websites describing the huge investment going into the town are silent on the steps being taken to protect the lakes and rivers and the wildlife that was still there in profusion when I visited. This will be Noyabr'sk all over again. Nor can I find any information on the fate of the Selkup community I visited, but it is hard to believe that they have not moved north yet again.

Krasnoselkup is turning out to be a major gas find, bigger than the existing main gas-producing area more than a hundred miles to the west, around Novy Urengoi. The two areas together make Gazprom probably the largest gas company in the world. And Gazprom doesn't let anyone forget its power. In Novy Urengoi, Gazprom chooses the mayor and runs the local TV station, which is called Gazprom TV. When I flew there in 2005, I was told by a burly policewoman at the airport that my Russian visa was not enough. To enter, I needed a special invitation supplied by Gazprom officials in the town.

As she filed my passport in a safe and got distracted by new arrivals, I wandered into town anyway. In the language of the local reindeer herding Nenets (close cousins of the Selkup), Urengoi means "rotten, godforsaken place." I could see what they meant. The wind howled among the bare apartment blocks. There were huge areas of sandy wasteland where dogs roamed, followed by men with vodka-red faces in leather jackets waving mobile phones. Women in high heels and fur coats clambered in and out of yellow Gazprom minibuses as fast as they could, to avoid the wind destroying their elaborate hairdos. The way into the half-built hotel was over wooden planks. The heating was off and the corridors smelled of paint.

Novy Urengoi provides a third of Europe's gas. The only foreigners in the place were German gas people, looking out for their nation's heating. If Gazprom turned off the taps here, Europe would get mighty cold in winter. And with new supplies soon coming from Krasnoselkup, Gazprom has big plans to replace Britain's diminishing North Sea reserves with Siberian gas. At the time of writing, Gazprom had already bought one British gas supply company, Pennine Natural Gas. And there were press reports that it wanted to take over Centrica, the owner of British Gas, the country's leading distributor.

Eventually they came to get me. A young spook and his interpreter had been detailed to round me up and take me back to the airport. They worked for Gazprom, of course. The spook wore an open double-breasted suit of cheap cloth to display his importance. I explained that I had come to see some scientists from Tomsk. He wondered how I knew that the scientists were not mafia, and would not kidnap me. I wondered how I knew he was not mafia, because he had certainly kidnapped me.

The airport departure lounge was full of construction workers in overcoats returning home after a weeklong shift in the "rotten, godforsaken place." The walls were covered with picture of Nenets, their camps, and their reindeer—celebrating a world that gas had destroyed.

My Electricity 20
Old King Coal Lives On at Drax

I wanted to know where my electricity came from. It proved a hard question to answer. My electricity comes, via a local supply company, from Britain's National Grid, which in turn gets it from generating companies that run power stations. But electricity is not a commodity like any other. It doesn't come in packets that I can track. It is an electrical current, maintained by the constant whirring of turbines at power stations all over the country—and one or two in France as well—and distributed by the National Grid.

Only the very biggest users can answer the question fully, because they need dedicated lines from power stations. So Rio Tinto's giant aluminum smelter on Anglesey gets its power largely from the Wylfa nuclear power plant close by, and will probably shut when the power plant closes in 2010. Likewise the huge surges of power required to run Europe's nuclear-fusion research reactor at Culham in Oxfordshire come down a high-tension cable from Didcot power station.

So who supplies the National Grid? It has three main types of power sources: nuclear, coal, and gas. The diminishing band of nuclear power stations were mostly built more than thirty years ago, usually in remote spots round the coast, in case anything went wrong. Many of them are far past their design lives and cannot be kept going much longer. Their fuel is uranium, mostly from Australia and Canada. Inside the power station's reactor, the metal is bombarded with neutrons to create heat that makes steam that drives turbines that generate about a fifth of our electricity.

It's clever stuff. Six grams of uranium fuel generate as much electricity as a ton of coal. My household benefits from the irradiation of less than a fifth of an ounce of uranium a year. The technology is rather old hat now. But the risks of getting it wrong are fearsome, as

Chernobyl showed. And we don't yet have any final resting place for the spent fuel, which is extremely radioactive. Currently it is stored in giant swimming pools at the power stations and occasionally goes by train, in high-tech and hopefully impregnable flasks, to Sellafield in Cumbria for storage. The government promises not to allow any more nuclear power stations to be built until it has figured out what to do with this accumulated waste.

Since the 1960s, Britain has been awash with cheap natural gas from beneath the North Sea. At first we used all of it to heat homes and run factories. Then it became a fuel in power stations, too, as the coal mines were shut. But now North Sea gas is running out. Increasingly, our gas is coming from abroad, in refrigerated tankers from Algeria and Qatar. Within a decade Britain may be importing 90 percent of its gas. Ten large gas storage projects are currently being developed to receive those imports. But meanwhile, international gas prices are rising. As a result, greenhouse-unfriendly coal, once written off as yesterday's fuel, is making an unexpected comeback.

In 2006, gas use in power stations shrank to about a quarter, while coal returned to the number-one spot, with a 40 percent market share. The remaining 15 percent or so is either imported from France (an extra dose of nuclear fuel) or is from renewables like wind and, predominantly Scottish, hydroelectricity. But for my footprint, the trail to follow is coal.

Anyone traveling north from London to Newcastle and Edinburgh will see the three giant power plants on the flat lowlands around the River Trent. They call this Megawatt Valley. It is among the most concentrated sources of electricity generation—and greenhouse gas emissions—on Earth. The three plants are named after the villages they invaded: Eggborough, Ferrybridge, and Drax. They were built by the Central Electricity Generating Board back in the 1960s and 1970s to burn coal from a planned new coalfield at the nearby Yorkshire town of Selby.

Drax was the last of the trio, completed in 1974. It is Western Europe's largest power station, with a capacity of 4,000 megawatts, so large that the company running it does nothing else. Drax produces 8 percent of Britain's electricity. Its main building is a quarter of a mile long and it has twelve cooling towers. Its 820-foot chimney was once Western Europe's biggest single source of acid rain. They cleaned that up by fitting it with "scrubbers" between 1988 and 1996 to absorb the sulfur emissions from the burning coal. This was Britain's initially grudging contribution to a continent-wide effort to take the acid out of Europe's rain. It was worth it. Since then, the fish have been returning to Scandinavian lakes.

But greenhouse gases? There are as yet no scrubbers to remove the carbon dioxide. Some 24 million tons of CO_2 goes up Drax's infamous chimney each year. That is as much carbon dioxide as comes from a quarter of Britain's cars. And it is more than all the planes leaving Heathrow airport produce in a year. Drax emits more carbon dioxide than Sweden. And it has precious few plans to do anything about it.

Under the EU emissions-trading scheme, which set levels for emissions from large industrial plants starting in 2008, Drax is allocated 10.6 million tons of CO_2 emissions, less than half what it actually intends to emit. The chief executive at Drax, Dorothy Thompson, plans to make up the difference by buying permits from companies with spare ones, including Centrica, the owner of British Gas. Centrica has spares because it can't sell as much gas to the National Grid as it would like to—partly because Drax is undercutting it. It's a funny old world.

One day, Drax may capture the carbon dioxide from its smokestacks—and perhaps liquefy and bury it in the empty oil wells beneath the North Sea. But Thompson is cool to the idea. "An interesting concept" is as far as she will go. Instead, she says she hopes to convert Drax to replacing 10 percent of its coal with rapeseed, willow coppice, and elephant grass grown by local farmers. That's the hope. But in 2006, Drax actually cut its biofuels use by 90 percent, because coal was cheaper. Who said King Coal had been dethroned?

This kind of short-term ultra-commercial approach does not win friends. In the summer of 2006, demonstrators camped outside Drax to protest against its emissions. They wanted to shut the place down. In that they failed. But they will return. I might join them next time.

One thing perplexed me. Drax burns 33,000 tons of coal a day.

But Margaret Thatcher shut down the Selby coalfield, so where does it get its coal? Half, it turns out, comes from British open-cast mining operations. The other 4.4 million tons comes from abroad. For all its claimed green credentials, Britain is today the world's fourth-largest coal importer—shipping in more than 50 million tons in 2006, double the figure for 2000. Megawatt Valley takes most of the 13 million tons entering through nearby ports on the east coast of England. Drax is among a group of power companies pushing to invest in port capacity so it can import more.

The biggest source of Britain's coal imports today is Russia. Most of the coal comes from beyond the Ural Mountains, from Kuznetsk in southwest Siberia. Dug from deep mines of the kind long since closed in Britain, it travels overland for 2,500 miles, starting on the Trans-Siberian Railway and then taking branch lines across European Russia all the way to the Baltic coast. From there, it goes by ship to Immingham, and then returns to rail for the last hop up Megawatt Valley to Drax and its fellows.

Kuznetsk seems a very distant place indeed. And it is. The journey is only possible because the Russian government provides huge subsidies for its mines and railways. Coal used to be regarded as too expensive to move very far. It was burned close to where it was mined. It was also owned by state enterprises and earmarked for keeping the home fires burning. No more. Coal has become a global commodity, and the big mining conglomerates have taken control. The largest, BHP Billiton, operates coal mines in Colombia and South Africa, Australia and New Mexico and Alaska. The other two kings of coal are Anglo American and the Swiss company Glencore.

These Big Three supply most of the rest of the coal that Britain imports. Thanks to them, South Africa is the world's third-largest coal producer and second-largest coal exporter, after Australia. Britain is among its biggest customers, taking more than 14 million tons a year. Almost all of it is mined in the Drakensberg mountains and exported from the Richards Bay coal terminal on the west coast north of Durban. The terminal exports around 77 million tons of coal a year, a figure that will probably have risen to 100 million tons by the time you read this. The United States is one of the world's largest burners of coal, which fuels half its power stations. Despite climate concerns, U.S. coal burning has been rising through the current decade and, with the country possibly having the world's largest coal reserves, could well continue. People talk about global oil production reaching a peak. Hooray. But coal production shows no sign of peaking.

Part Six
Downstream

My Rubbish

Down the River
and across the World

21

My waste goes downstream. Sometimes a long way.

Once, I interviewed a woman who literally produced no rubbish at all. She recycled or reused everything. She didn't even possess a trash can. I'm still not sure how she did it. The average Briton produces more than a thousand pounds of household waste a year and only sends a quarter of that for recycling. Which sounds like a lot but is small compared to the nearly 1,800 pounds produced by each American, of which almost a third is recycled. Every Wednesday morning a garbage truck comes to empty my trash can. It takes the contents a mile or so downhill from my house, past fast-food restaurants, home goods stores, and a budget hotel, to the Western Riverside Transfer Station, close to where the River Wandle joins the Thames. One morning I followed.

The transfer station is a whopper, as tall as a church tower and with a central aisle—where the garbage trucks drive and dump their contents down twelve chutes into containers—that is longer than a large cathedral. No wonder that Lynne Cure from Cory, the company that runs it, calls it a "cathedral of rubbish." It has loomed on my skyline for years but I had never been inside. It takes the contents of something like a million trash cans from four London boroughs, delivered by relays of trucks driving up a ramp one at a time, at the command of a green light. The trucks are almost lost inside, as they lift up their rears and dump their 13-ton loads. My Wandsworth truck emptied my rubbish down chute number six.

This transfer station has always been controversial. When it was first proposed twenty-five years ago, there were protests, and I remember we were promised it would only be around for a few years. Nothing permanent. Lynne says that was never true. And Cory now has a contract to handle waste there till 2032. Rubbish is a bad neighbor. And like most bad neighbors, it is hard to dislodge.

We left the cathedral and went around to the river side of the operation, where the chutes empty into containers. Out there, even the containers looked small. A huge crane picked up the container with my rubbish and lifted it onto a barge floating at the dockside. Each container holds nearly 14 tons of rubbish. Each barge can take thirty containers. So that is more than 400 tons. And the tug that would take us downstream, the *Regain,* was lining up to pull three of the barges. Some days it takes four. With three journeys a day, they can take nearly 5,000 tons of rubbish in a day. From just four boroughs. We hopped aboard the *Regain,* waited for the tide to turn, and we were off. Me and my rubbish—and that of a few tens of thousands of other people—heading downstream. Always downstream.

The *Regain* wasn't pulling all the rubbish my household had produced that week. There was also the recycling collected the same day as the bins were emptied. The small municipal truck we saw crossing Wandsworth Bridge as we headed downstream could have contained my recyling. It was also headed downstream. The council website says that its recycling is "sent directly to reprocessors for recycling into new products." Up to a point, as we shall see later.

On board the *Regain,* we made a steady 4 knots. Before we were under Wandsworth Bridge the cook had offered us ham, eggs, and fries, which we ate in front of a centerfold of a model with her legs apart. Being Thames lightermen, as these truckers of the river are called, is a time-honored trade. It takes longer to learn the tricks of the river than it takes to learn "the Knowledge" to be a black-cab driver. Or it did. New EU rules now allow lightermen trained on the Rhine and Danube to ply the Thames with just six months' preparation. This is not appreciated down on the tideway. Not least because the Rhine bargemen could realize that, strictly speaking, these tugs might not need a crew of five (Captain Geoff plus the bloke who steers, two crewmen, and the cook).

Soon we were past the shell of Battersea Power Station, the flash flats of Chelsea Harbor, the Houses of Parliament in Westminster, St. Paul's, and Tower Bridge. There is no better way of seeing the sights of London than by cruising down the river, even if you have several hundred tons of putrefying rubbish on your tail. Soon we took our precious cargo past Canary Wharf, through the Thames Barrier and past the jetties of Tate & Lyle, the world's largest sugar refiners, where the bulk carrier the *Buse Stevens* was tied up. Our barges still had a way to go, but they stayed overnight just downstream of the Barrier. Our two ropemen finally earned their egg and chips, leaping from the tug onto a pontoon and securing the cargo for the night.

I haven't yet mentioned my third waste stream, which was also coming downriver. My sewage. London's geography used to be full of references to its crude waste disposal systems. The capital's foremost biographer, Peter Ackroyd, records Pissing Alley and Dunghill Lane, Midden Lane and Shiteburn Lane. Until little more than a century ago, the city was full of night-soil men, scooping up its ordure and removing it in the dead of night.

Dickens gave one dump site pride of place in *Our Mutual Friend*. But most waste never reached anything so formal. Filth was everywhere. The city's trades included bone pickers and rag gatherers, cigarette pickers and sweeps and dredgermen and mudlarks—all engaged in the ancient rites of recycling the city's discarded rubbish. After the water closet was invented, more of the city's effluent went down the sewers, where freelance scavengers known as toshers were soon lifting manhole covers and sliding down the pipes to see what they could find.

The sewers emptied into the river, and by the mid-nineteenth century, the stench on the Thames was so bad that Parliament hung sheets soaked with chlorine across its windows to ward off the worst. When such measures ceased to staunch the stench, they paid an engineer called Joseph Bazalgette to construct two giant sewers, one along either bank of the Thames, to take the capital's excrement downstream. The Victoria and Albert Embankments are built on top of them. The sewers mostly contain what you would expect. But there are also pharmaceuticals and other small toxic things that seem easiest to get rid of by flushing. Then there is paper. A lot of it. Each one of us apparently uses more than eleven thousand sheets, or close to 30 pounds, a year. That's thirty sheets a day. (No, I don't either. But that's the official figure.) For 7 million Londoners, that is nearly 100,000 tons of toilet paper down the toilet a year, or a million trees.

The north-bank sewer ends at Beckton in east London, the largest sewage-treatment works in Europe. My sewage, making its way along the south bank beneath Southwark Cathedral and the Globe Theatre, ends up at the only marginally smaller Crossness, just downstream of the mooring point for the barges. Read Thames Water's PR and it sounds like a rural idyll. It is surrounded by one of the last areas of grazing marshlands in London. There are nesting sand martins, reed beds, a bat cave, and even cycle access. But, despite the wildlife, this is a pretty typical treatment plant, where some 185 million gallons of sewage are treated every day, enough to fill the Albert Hall seven times over.

The sewage is filtered to remove large items, then mixed with bacteria that eat up organic matter, and allowed to settle. The liquid is discharged into the Thames estuary. And the solids are left to dry to a thick sludge. Londoners like me individually produce around 45 pounds of sewage sludge a year, or 2 ounces a day. One-third of the sludge is burned in two east London incinerators. The other two-thirds is sold to farmers as a cheap fertilizer. Yes, the night-soil men are still in business, only now wearing Thames Water overalls.

Downstream from Crossness we reached Crayford Marshes, home of a "materials recovery facility" used by Wandsworth and many other London boroughs to handle their domestic recycling. Here a company called Grosvenor sorts 440 billion tons of miscellaneous recycled paper, cans, glass, wood, and plastic a year. The sorting begins at a spinning drum with holes in the walls. As it spins, bottles and cans tumble out, while paper and cardboard stay inside. Then the rubbish goes along conveyors where magnets hook out ferrous cans and jets of air remove light plastic bottles.

The Grosvenor plant, and others like it, do the jobs we are too lazy to do, separating our recyclables into different streams. It even divides up paper so that the newsprint recyclers don't get stuck with cardboard and Yellow Pages and pink *Financial Times* that would muck up their processes. After getting sorted at Grosvenor, a lot of London waste does get recycled, mine included. But what surprised me was how and where—and that many of the claims being made for what happens to our recycling are simply untrue.

Take glass. It makes up around a quarter of our household recyclables. The notice on the bins accepting glass outside my local supermarket declares that recycling one glass bottle or jar saves the energy "to power a TV for 20 minutes." Well it might if it were turned into new glass, but it isn't. Instead my old bottles go to Day Aggregates in Brentford, west London, where they are ground up and mixed with the 3.3 million tons of construction materials the company delivers across southeast England each year.

There is a certain logic to this. Glass started out as sand and now ends up as, in effect, a substitute for sand. But is turning rubbish into aggregate really recycling? Now that local councils have targets for recycling, this is an issue over which lawyers are doing as much fighting as waste technologists. And I am not too happy with my recycled glass going to make new roads. I expect many greens would be similarly upset.

My aluminum cans go to Novelis on Merseyside and my steel cans to AMG Resources, which strips the tin coating off for sale as tin ingots, and adds the steel to the huge international trade in general metal scrap that is eventually smelted and turned into new goods. It might go to the steelworks of Bangladesh, for instance; scrap metal is Britain's biggest export to that country.

Britain produces 3.3 million tons of plastic waste a year (roughly a tenth of the output of the United States), most of which ends up in landfill. But we are getting better at handing plastic on for recycling, especially plastic bottles made of PET (polyethylene terephthalate) and HDPE (high-density polyethylene), and there is a ready international market. Grosvenor says its PET makes fleece and car upholstery and its HDPE often resurfaces as drainpipes. For now, roughly half our plastic bottles returned for recycling are sold to China. Why? It is a mixture of cost and bureaucratic inertia.

Ray Georgeson at the government-funded Waste and Resources Action Programme (WRAP) says a British plastics-recycling business could be kick-started if plastic bottle makers would agree to accept 25 percent recycled feedstock. But they won't. The going rate for old plastic bottles is about $240 a ton, which is about half a penny per bottle and more than twice what British recyclers say they can afford to pay because of their higher wages.

Recycling of plastic still seems to make sense in energy terms, even if it is done in China. Georgeson says that every ton of plastic waste that is recycled prevents the emissions of about 2 tons of carbon dioxide in making new plastic. Certainly the case for recycling plastics looks more robust than that for recycling paper right now.

Britons send more than 7.7 million tons of paper a year for recycling, or 265 pounds each. Americans recycle more than 48 million tons of paper annually, or 330 pounds each. We have a simple faith that this is a good idea. The notice on my local recycling bin says with disarming simplicity: "Each tonne of paper recycled saves an average of 15 trees." But the closer you look, the harder it is to be sure. The biggest UK market for newspapers and magazines is the Aylesford paper mill in Kent. Grosvenor also sends my cardboard and "mixed paper waste" to the Smurfit Mill in Kent, another giant of the industry. And my old office paper goes to M-Real's plant for making the Evolve brand of new office paper.

It sounds like almost 100 percent domestic recycling. But it doesn't always happen like that. Most Western countries—the United States as much as Europe—find that it is cheaper to export their waste for recycling, partly because it is a fairly labor-intensive activity and partly because the paper mills that could use the material are closing. In Britain, more than 4 million tons of wastepaper is exported, and the amount recycled at home is declining, even though the amount we collect for recycling is soaring. For companies like Grosvenor, the pressure has been on to sort and find homes for all this lovingly collected material. And in Europe there is an additional issue: exporting paper waste for recycling is legal and encouraged, whereas exporting the same paper for disposal is illegal. Everything depends on the definition, and things have been going wrong.

Poor old Grosvenor. At the end of 2005, the BBC discovered a container load of paper waste from London boroughs on a dockside in Jakarta, Indonesia. Inside were private letters tossed in bins months before by the citizens of Islington. There could have been my bank statements and credit card receipts, for all I know. The reporters brought some of the letters home and popped round to ask what their previous owners thought about Islington's recycling practices.

The fact of our paper waste being exported is not exactly news. What was odd was the route taken by this particular cargo, as eventually tracked down by the council. The 495 tons of wastepaper went in twenty containers from Crayford by truck to a "materials processing" plant in Münster, Germany. But the company rejected the containers in January 2005 because it said they contained too much "contamination." So Grosvenor found another buyer, a mysterious "reprocessing company" on the other side of the world in Jakarta. In early February, Grosvenor consigned the containers via Rotterdam. Then there is a long gap till June. By this time, the mysterious Jakarta company had equally mysteriously gone out of business. The BBC wondered aloud whether the company had ever existed, since the address on the shipping documents was a closed-down Japanese restaurant. At any rate, the consignment was left sitting on Tanjung Priok dock.

After the BBC got a tip-off and went filming, the container loads were hurriedly sold for a third time, to a company in Malaysia. After turning up at the Malaysian port of Port Kelang in August, they appear to have been sold one more time, before finally reaching a bona fide recycling plant in the capital city of Kuala Lumpur in October, where the contents were eventually turned to pulp around the end of the year and sold to local paper mills. It probably turned up eventually in the *New Straits Times*.

The entire journey from London to Kuala Lumpur took eleven months and involved a host of changes in ownership and legal jurisdictions. The paper seems to have been "sorted" at least three times. The journey was not typical, but it was symptomatic of a dysfunctional industry. And it formed part of a pattern at Grosvenor that led the Environment Agency into a series of investigations, culminating in a conviction in April 2007, when Grosvenor pleaded guilty to the export of the equivalent of ninety truckloads, more than 4,400 tons, of unsorted household waste found in containers opened during a crackdown on the dockside at Rotterdam. The paperwork said the containers were full of "mixed card and paper" destined for recycling in China. In fact, according to the Dutch authorities who impounded them, they also contained plastic packaging, batteries, drink cans, old clothes, grocery bags, and wood.

Europe bans the export of waste but positively encourages the export of material for recycling. Unfortunately the boundary between the two is often blurred. Grosvenor said the contamination wouldn't stop the paper and cardboard from being recycled. But the Rotterdam authorities and the Environment Agency disagreed, and Grosvenor faced charges.

Even as the Islington papers were en route from Rotterdam to Jakarta, the head of waste management at DEFRA, the government department responsible for overseeing the recycling business, had sent out a circular to local councils warning of widespread "illegal exports of municipal waste." After some spot checks, the Environment Agency concluded that half the containers leaving Britain labeled as carrying wastepaper for recycling abroad contained no such thing. Grosvenor seems only to have been unlucky in being found out.

Every year, Britain consumes more than 13 million tons of paper and board (440 pounds per head), which requires 150 million trees. To grow that amount of timber each year would take a forest the size of Wales. The United States consumes almost ten times as much (or twice that much per head). On the face of it, the argument for recycling our paper and cardboard is irrefutable. And it is hard to argue with the sheer scale of the operation carried out just down the road from the Grosvenor plant in the village of Aylesford. Here on the River Medway, a traditional British paper-making center, is Europe's largest paper-recycling mill. It is entirely devoted to manufacturing newsprint for our daily papers—from old newspapers and magazines, including mine.

Aylesford receives one in five of all the newspapers and magazines recycled in Britain. Some thirty thousand truckloads of paper come through the gates each year. That's more than 70 million newspapers, the product of more than 5 million trees. The paper is soaked in water to make pulp, treated with soap and solvents to remove ink, and then screened, spun, and further treated to remove staples, plastics, glue, and grit. Cleaned-up pulp is finally re-formed into paper on colossal machines that turn out a constant stream of newsprint nine yards wide at a speed of more than 60 miles an hour. Almost one in every twenty newspapers in Europe is made from newsprint from this one plant. On average, they say, the complete cycle from mill to printing press, newsstand, breakfast table, recycling bin, and back to the mill takes about fourteen days. The biggest leakage in the closing of the loop comes from freesheets (free daily newspapers handed out on the streets), most of which end up in municipal bins or picked up by the people cleaning buses and trains. Mostly those end up in landfill.

But despite the plant's obvious success, some are questioning whether the routine recycling of newspapers is environmentally the best option. I first came across this a decade ago when I wrote a cover article for *New Scientist* titled "Burn This." It suggested that so much energy was being used carting wastepaper around the country for recycling that it would be better to incinerate it. One study in rural Norfolk found that cars traveled more than 165 miles in journeys to local recycling points for every ton of paper waste posted into the collecting bins. And even once the paper reaches the bins, it probably has a long journey to Aylesford. On the other hand, incinerators are usually more local, so the energy for trucking wastepaper to the incinerator is much less than trucking it to a paper mill. Mills create new paper, but incinerators can create large amounts of energy.

Of course if you burn the old paper, you have to replace it. And it still takes more energy to make virgin paper than to recycle old paper. So recycling is still better. Or it would be, except that most of our virgin paper comes from Scandinavian mills, where they get their energy from burning wood chips that are constantly being regrown in the forest. The mills are effectively carbon neutral. Take that into account, says Matthew Leach of Imperial College London, and making virgin paper consumes roughly half as much energy as recycling old paper. So burning is best. Sacrilege? Maybe, and other studies have reached different conclusions. But it is certainly interesting.

I know that lots of people hate incinerators, fearing dioxins coming out of the chimneys. My reading is that a lot of old municipal incinerators did indeed emit toxic nasties, including dioxins, and there is some evidence of cancer clusters around some of them. But modern plants are much cleaner. And, if they also generate "green" electricity from my waste, then I am in favor of them.

So much for recycling. But what of my third waste stream—my sewage? I last saw this at Crossness, where the solid was sold to farmers and the liquid was deemed clean enough to discharge into the outer reaches of the Thames estuary. Thanks to a mixture of European directives and public outrage there has been a major effort in recent years to clean up our rivers, estuaries, and seaside beaches. Things aren't perfect, but from a public health perspective, they are a lot better. But here is an unexpected problem. Nature finds our rivers and estuaries and coastal waters just too clean.

On my journey down the Thames, where did I see most wildlife? It was at outfalls of sewage and drainage water, where birds gathered. Birds also followed the waste barges, collected at the Mucking landfill site, and pecked at the litter-strewn "strandlines" along the water's edge, where rubbish washed up. The birds don't want the sewage or the rubbish so much as the other creatures that thrive on it. And worms, the estuary's famous cockles, and other invertebrate species also like our rubbish.

John Pethick, a leading coastal geographer, first pointed this out to me. As we have tidied up our fields, conserved soil on the land, and concreted over our urban areas, rivers have become starved of natural organic matter from soils and rotting vegetation. For a long time that natural organic matter was replaced by the huge amounts of sewage effluent we put into the rivers. Too much untreated sewage killed the river, of course. But partially treated sewage effluent, especially when discharged into estuaries and bays, often acted as a convenient fertilizer for river plants and sustained their ecosystems. "Where treated sewage is still discharged, it is often about the only source of nutrients left in our rivers," says John. "And when that goes, wildlife suffers."

Mark Rehfish at the British Trust for Ornithology agrees. "We have seen dramatic declines in wading birds round our shores. The loss of sewage discharges is a major factor." Twenty thousand wading birds disappeared from Liverpool Bay after a sewage outfall on the Mersey estuary was shut in the 1980s. One of Western Europe's largest populations of sandpipers disappeared when the sewage from another town was cleaned up. New sewage works in northeast England had a similar effect on birds theoretically protected by a European directive on wild birds. Arguably that makes the sewage works, required under one EU law, illegal under another. The Thames estuary too has been starved of nutrients.

It is a startling discovery: Cleanliness is not necessarily an ecological virtue. Often, it is a vice.

Trade Not Aid

22

Joining the Great
Global Rummage Sale

The Milonge brothers were on a roll. They came to pick me up at my hotel in downtown Dar es Salaam, eager to show off their business selling the Western world's cast-off clothes. In Tanzania, they call old shorts and shirts and skirts and socks *mitumba*, meaning a bale. That's how they come, in bales unloaded from containers at the Dar dockside just down the road.

The delivery of mitumba to Tanzania started as charity, but now it is big business. Mitumba is the biggest export to Tanzania from both Europe and North America. The European Union alone exports more than 11,000 tons, valued at about one euro, or $1.20, per kilogram (about 2 pounds). And Geoffrey Milonge and his brother Boniface are proud of what it has done for them since they came to Dar from their village out in the bush ten years ago. We piled into the big car, and I couldn't help noticing their big rings, big stomachs, and, to be fair, big smiles too. The brothers are generous and keen to do business.

On average, each of us buys around 75 pounds of textiles a year. We eventually throw about 65 pounds of that into landfills and hand over about 10 pounds to charities such as Planet Aid, Oxfam, and the Salvation Army. We probably don't think much more about it, beyond having a vague sense that the clothes are probably given to someone who needs them. So after putting a bag of old clothes into the local bin operated by Scope, a charity that helps people with cerebral palsy, I set out to discover what happened to it.

Scope told me that the contents of the charity's recycling bins are emptied weekly by contractors and trucked to a depot. There, the clothes sit for a while in a big pile, until smaller vans turn up and collect bundles of stock that are taken to individual shops. I knew from a friend who works in one of the shops that there are not enough UK buyers for even the relatively small amounts of old clothing that people like me put in the bins. I do my bit. My friend brings me near-new, top-brand shirts that have been clogging up the racks in Scope shops in south London because nobody wants to buy them, even for a few pence. But with the malls full of cheap clothes, I seem to be in a small minority.

So plenty of unsold stock is left "out the back," as Scope's recycling manager, Nick Wilks, put it, for recollection, this time by rag merchants. They sort the clothes again and decide what they can sell where. It could be resale to people in Eastern Europe or Africa or for industrial rags.

This struck me as a rather convoluted and wasteful distribution system, in some cases involving clothes being driven across the country several times. But I had stumbled on a sizable trade, a modern version of the old rag-and-bone merchants. One of the firms that buys from Scope, JMP Wilcox handles 560 tons of old textiles a week and fills seven or eight containers for export. Wilcox advertises bales in forty-five categories, ranging from ties and net curtains to men's zipper jackets and women's bras. I contacted a few of Scope's regular buyers. One trader, Lynne Wright, who runs a company called Wastesavers, got in touch. She had for several years been shipping container loads of old skirts, blouses, and much else to a man called Kazim Remtulla, whose splendidly named Mohammed Unique Company is based in Dubai. Not that the people of Dubai wanted the clothes. They were en route to Africa, mostly to Africa's largest mitumba markets in the streets of Dar es Salaam.

Kazim had recently pulled out of his arrangement with Lynne. The weak dollar meant he could do better buying from the United States, she said. But Lynne had a new customer, a mitumba trader in Dar who had been buying from Kazim and wanted to do his own trading. And that is how I came across the Milonge brothers. Geoffrey responded to my e-mail within hours. To be honest, his poor English had led him to think that I might be planning to sell him some bales of good-quality British mitumba. He pointed at my shirt. "We'd sell that for four thousand shillings [about $3.00]." But even when he learned I was only a journalist, he kept up the bonhomie. Great guy. Good businessman.

We were in the car in seconds and heading downtown. The brothers told me they were excited about receiving their first consignment direct from abroad. They were cutting out the wahindi—their name for the Asian middlemen who had ruled their lives so far. Lynn's container had five hundred bales, each weighing roughly 100 pounds. It was still at sea, probably somewhere in the Suez Canal, but it would be arriving in Dar port within a week.

Soon we were in the heart of the Kariakoo district of Dar, a big central trading area. The brothers had a corner shop at the junction of Kongo and Mhonda streets. It was full of men and women picking through the piles of old clothes. Showman that he was, Geoffrey had his knife out within seconds. He had just bought a new bale from the port, dispatched by another UK supplier. The label said "LMB London"—which turned out to be Lawrence M. Barry & Co. of London. LMB too handles textile hand-me-downs from Scope. The bale contained 240 men's shirts, for which the brothers had paid 200,000 shillings. At $165, that was double the usual price for British mitumba. It worked out to 67 cents a shirt. But men's shirts are the biggest seller in Dar, and Geoffrey reckoned there was a gold mine here. What would they sell for? I watched.

It was a grab bag. There was a Paul Smith shirt in there; it sold quickly for 7,000 shillings, or $5.80—a markup of more than 800 percent. The female buyer said she planned it as a present for her lover. An almost new brown M&S shirt went for 5,000 shillings to a middle-aged man. A shirt from either of two typical British clothing stores, Next and Burton, would sell for nearly as much, Geoffrey reckoned. Several customers told me they were looking for fashionable labels, even though they would pay more. In Dar as much as in Oxford Street, labels sold.

Geoffrey reckoned overall to make a 20 percent profit, but it depended on how lucky they were with the bales. A good bale might fetch more than a million shillings and yield 500 percent. "The best quality is from the UK," he told me. "English clothes are much less worn. The U.S. is good too, but their clothes are rather large. The Germans wear their clothes too much." You learn a lot about a nation from its mitumba.

The next morning the brothers had the car at the hotel again. We went to some of their market stalls out in the suburbs. At the Mukumbusho market, there was a huge pile of American clothes all for sale at 500 shillings. A John Henry red dress caught my eye. Another pile of trousers retailed at 1,500 shillings each. Here were some brown cords very like the ones I was wearing. Made in Mexico. There were Land's End jeans made in Malaysia and Cherokee chinos from Hong Kong and Mauritius, as well as Bonjour bags from the Dominican Republic and Haggar and St. John's Bay trousers from Mexico.

The brothers sold fifty bales a month from this stall alone. "It's all American at present, because that's what the wahindi are buying. But Americans are so big, we have lots of clothes left over," said the guy running the stall. Geoffrey could have worn his stock okay, but his thinner compatriots would have been swamped by many of the mega-girth jeans and belly-stretched shirts I saw on display. But Lynn's British container was at sea. All would change soon.

Next was the Urafiki market. Three-quarters of Dar's food is traded here. All around are cafés and photo studios, grocery stores and furniture emporia, butchers and beauty salons. And the Milonge brothers had one of the best spots. They sold more variety here. I saw branded American leather hiking boots for 40,000 shillings; Adidas soccer boots for 15,000; Reebok pumps for 5,000; women's business suits and secondhand Korean schoolbags for 2,000; and Canadian baseball caps for 1,500. Bras were popular at 2,000; slips at 1,500.

Here I also found some new clothes on sale. They were mostly end-of-line stock from U.S. budget stores. Some still had their final dollars-and-cents price tags on. I saw some Gap Oxford-weave shirts, made in Indonesia, and Van Heusen shirts made in Thailand, both with thrift store sale stickers for $0.95, but on sale here for 2,500 shillings. That was almost $2. So these clothes were being sold in Dar at more than their final knock-down price in Denver or Des Moines.

The mitumba trade shows how Africa has changed. When Tanzania became independent in 1961, one of the first acts of the first president, Julius Nyerere, was to ban imports of used clothes. Apart from disliking the heavy symbolism of wearing colonial castoffs, he wanted to build up a domestic textiles industry. Sadly, the new industry, backed initially by state subsidies, attempted to ape Western tailors rather than providing cheap and cheerful clothing for the masses. Result: millions of men wearing ill-fitting suits and women in dowdy skirts and blouses.

As Africa got poorer and trade globalized, the indigenous textiles companies collapsed and mitumba started to creep in again. First, it came through the back door, smuggled from Mozambique and Kenya. Then Western charities started handing out old clothes for free. But demand was so great that soon they were charging, to raise money for other projects. And then the business got handed over to the wahindi and other entrepreneurs. The charities restricted their activities to collecting the clothing back in Europe and North America.

Some people in rich countries are surprised that clothes they donate to charities end up being sold to poor people for a profit. But this is "trade not aid," and the Milonge brothers are doing very nicely, thank you. In her book *The Travels of a T-Shirt in the Global Economy*, the American economist Pietra Rivoli said the mitumba business was the nearest thing to a free market in the heavily subsidized and quota-controlled cotton industry. That may not be a virtue in itself. But supply does meet demand in the markets of Dar. At least a third of Africans dress in our old clothes, and in Tanzania the figure rises to 80 percent. As many visitors have commented, Tanzanians do seem to be much better dressed than they used to be.

Beyond the Grave   23
A New Life for Joe's Old Phone

Our household survived without a mobile phone for a long time. Even as a journalist, I resisted. The first arrived when our son Joe bought a standard Nokia in 2003. Always on the lookout for a bargain, he paid £10 (about $20) at Carphone Warehouse. It did the job. Or it did—and there is no sensible way of writing this chapter without sharing the family tragedy with you   until his death in 2005. In fact it did the worst job of all. After Joe collapsed while out jogging at university, the nurse at Leeds Infirmary found "MUM" in the memory and used it to tell us the news. Anyhow, this is the story of that phone.

Along with the toothbrush, the mobile phone is about the most ubiquitous piece of personal equipment on the planet today. No invention has ever spread around the world to so many people so fast. Twenty new users are connected every second. Sometime in 2006, the two billionth phone went live. By the time you read this, there will be one for every three people on the planet. Even in the poorest countries, mobile phones are everywhere. In 2006, in Tanzania, where average annual income is just $560 a head, I saw more signs advertising refill cards from local phone companies like Vodacom and Celtel than billboards for Coca-Cola. Every village store, every roadside shack seemed to be selling them. And millions of Tanzanians were buying. The cheapest refill cost the equivalent of about 20 cents, enough for a 45-second call.

Joe's purchase plugged him into the modern world of global commerce like no other commodity can. Nokia is a Finnish company that began life 140 years ago making paper, before getting into rubber and finally electronics. Today it is the largest supplier of mobile phones in the world, with more than a third of the market. Nokia and subcontractors like the Taiwanese company Foxconn assemble more than two phones every second in each of nineteen factories spread round the world. Nokia says it has a strict policy of not revealing which factory individual phones come from. But it is most likely Joe's came from China, which makes roughly half the world's phones, and probably from Foxconn's plant in Shenzhen, the manufacturing boomtown in southern China, which employs seventy thousand people.

But these factories are just the end of a manufacturing network similar to that making computers. Every Nokia phone contains some four hundred different components. Each year, Nokia and its subcontractors assemble an estimated 100 billion resisters, capacitors, motherboards, display screens, keypad buttons, batteries, covers, and other parts. The Shenzhen plant feeds off dozens of local component manufacturers, like Delta, which employs twenty thousand young workers in nearby Dongguan and is the world's largest manufacturer of adaptors and capacitors for computers and mobile phones. And Hua Tong in nearby Huizhou, which makes Nokia's circuit boards. Like Foxconn, both are Taiwanese companies.

Shenzhen is a city literally made by mobile phones. Twenty-five years ago it was a fishing village surrounded by rice paddies. Today it is an urban sprawl of 12 million people—twice the size of Hong Kong—and stretches for more than 60 miles along the east bank of the Pearl River to Dongguan. I met a man who had been resident for eighteen years in Shenzhen. He claimed he knew nobody who had been there longer. I certainly felt like, and quite possibly was, the oldest person in the city.

But this is no bargain-basement metropolis. The city's gleaming new central business district has five-star hotels, a stock exchange, concert hall, huge exhibition center—and a profusion of Starbucks. This is the new China. Not just the workshop of the world, but increasingly a consumer capital too. They say that as many mobile phones are sold in the stores of Shenzhen as are manufactured in its factories. I can believe it. At one store I saw, at a rough estimate, ten thousand phones on display. The aisles were packed with buyers —not just for Western brands like Nokia, but for would-be Chinese global brands like CETCs, from the state-owned China Electronics

Technology Group Corporation. Many people were also buying parts and having phones custom built on the premises.

Companies don't make phones, of course; people do. And Foxconn and the other mobile phone manufacturers in southern China employ hundreds of thousands of young women migrant workers. Communist social control and capitalist money have combined here to extraordinary effect, creating the modern workshop of the world—as dominant in its influence as Britain once was at the start of the industrial revolution two hundred years ago. Somewhere about 2003, the delicate fingers of a young woman called Hua or Lee or Ling probably put together Joe's phone and shoved it in a box for dispatch to Clapham Junction.

Behind the component suppliers lies another vast industry, mining and refining the metals and other materials that go into the components. The phone is a miniature smorgasbord of metals and other ingredients. Many are toxic and accumulate in the food chain. They can damage the human nervous and reproductive systems, cause cancer, and damage our genes. An individual phone won't do this, but as the ingredients accumulate in the environment, they may. Nokia has won praise from Greenpeace for cutting the use of some of the nasties in its new handsets. But older phones, like Joe's, still contain them.

About half the weight of a typical mobile phone is the casing, which is made of molded plastic with a little added iron and aluminum. The plastic contains phthalates, which are often used to soften it, making PVC flexible, for instance, and also creating the "jelly rubber" in sex toys, I am told. In high doses phthalates, which have been dubbed "gender-bending" chemicals, damage hormonal systems. They shrink testes—which sounds like a bit of a turnoff for a sex toy. Another quarter of the phone's weight is wiring and the circuit board. These are mostly made of copper but also contain magnesium, tin solder, and small amounts of gold, as well as arsenic, chromium, and beryllium, which can produce toxic dust during manufacture and recycling. The circuit board also probably contains brominated flame retardants, which prevent your phone from bursting into flames, but may produce dioxins if the phone is one day incinerated. The phone screen is made of glass and ceramics. Then there are cobalt, lithium, and carbon in the battery, silver in the keyboard, and tantalum in the capacitors.

A typical mobile phone today weighs only around 2.5 ounces, but taking its many ingredients from the earth requires the mining of 65 pounds of rock. In addition, manufacturing the chips requires several hundred quarts of water, and the energy that probably comes from burning dozens of pounds of fossil fuels. And making the batteries and keeping them charged through a phone's typical two-year life raises the weight of a phone's overall materials rucksack to about 165 pounds—a thousand times the weight you carry in your pocket.

But what of the phone's footprint in human lives? A year or so before Joe's Nokia rolled off the production line in Shenzhen, the papers were full of stories about a mysterious substance called coltan that was essential to mobile phones. It was mined under inhuman conditions from the forests of central Africa, in the midst of a vicious civil war that had killed up to 4 million people. In pursuit of the footprint of Joe's phone, I decided to investigate.

Coltan turns out to be an African nickname for a bunch of minerals, usually found together, that contain two elements, columbium and tantalum. Tantalum is what the fuss was really about. Even rich Westerners get through less than one-tenth of an ounce of the stuff in a year, but tantalum is literally essential to the modern mobile phone. The hard, dark-gray metal has an exceptionally high melting point (almost 5,500°F) and is extremely resistant to corrosion and chemical attack. It can be pulled and pushed and squeezed to make tiny wires and sheets and tubes. And, most handily of all, it can be made to store and release an electrical charge.

Tantalum was first used as the filament in light bulbs, till it was replaced by tungsten. Then a bit over half a century ago, the military started using it to make capacitors in radar receivers and missile guidance systems. Capacitors are the tiny components of electrical circuits that store and release electrical charge. For a long time, most of the world's tantalum was mined in Western Australia by a company with the improbable name of Sons of Gwalia and sold to Cabot, a U.S. refiner, before being stockpiled by the U.S. military, which regarded it as a strategic metal. Another refiner was H. C. Starck in Germany, a subsidiary of the industrial giant Bayer. Tantalum has found other uses. It turns up in heart pacemakers and insulin injectors and automatic braking systems in cars. But the big boom came with rising demand for portable electronic devices like laptop computers, Play-Stations, digital cameras, and, most of all, mobile phones. You can make capacitors with ceramics or aluminum. But tantalum is more compact and operates at a low voltage. So batteries last longer. Without it, mobile phones would still be the size of bricks and need recharging every fifteen minutes. And the smarter the phone, the greater the number of tantalum capacitors. A phone with a camera and video functions requires more than twenty.

Today more than half the world's tantalum goes into making more than 20 billion capacitors every year—over three for every person on the planet. The world's biggest tantalum capacitor manufacturer is Kemet Electronics in the United States, which has a long-term supply agreement with Cabot. But when the mobile phone business took off, other tantalum refineries and capacitor manufacturers set up in Japan, Taiwan, China, and elsewhere in Asia. In 2004, Kemet opened a new manufacturing plant in Suzhou, the Chinese computer manufacturing center west of Shanghai (see chapter 15). Another maker is Guangdong Kexin in Shenzhen. Most likely one of these supplied the capacitors in Joe's phone.

As the world got hooked on mobile phones, tantalum demand soared. Speculators cashed in and tantalum prices rose fivefold in just a few months, to about $500 a pound. The rush was on to find new supplies. The first to be exploited were the coltan reserves in eastern Congo. By some estimates, the provinces of Kivu and Ituri contain 80 percent of the world's tantalum reserves. Mining it is easy: you make a clearing in the forest, dig a pit, and excavate the dull gray rock just beneath the surface. In those provinces some unsavory rebel military leaders were already conducting a vicious civil war. When coltan prices soared, they realized they could make fortunes out of the ore, and use it to raise cash for arms, or simply to fill their Swiss bank accounts.

The warlords overran some legitimate mines, while others remained open in circumstances that raised eyebrows. They also press-ganged child soldiers and prisoners to mine the coltan and ship it off to brokers, who were often little more than gunrunners paid in coltan. The UN Security Council began an investigation. It found that for several years, one or more Western mining companies were extracting coltan from eastern Congo under the protection of a rebel army. Much of this material, it is claimed, ended up with specialist companies like Starck.

The UN also reported that foreign armies invading from neighboring Uganda and Rwanda organized shipments. Top dog was Rwandan Brigadier-General James Kabarebe, who for a while made $20 million a month working with a mystery woman, dubbed Madame Coltan, who warehoused coltan at her cigarette factory in the eastern Congolese town of Bukavu. And it uncovered an alleged coltan trail to Eagle Wings Resources International, a subsidiary of an Ohio-based company called Trinitech International, through which other firms "obtained their own mining sites and conscripted their own workers to exploit the sites under severe conditions." Eagle Wings, it claimed, had "close ties with the Rwandan regime" and flew its coltan to Starck as well as to metallurgical plants in Kazakhstan and China. Starck has repeatedly denied obtaining coltan from central Africa and says it took steps to ensure that it did not do so. And the Chinese plant denied doing business with "any individual or any entity that represents somebody or some entity in the DRC." The investigating panel, while noting these denials in its report, claimed to possess information to the contrary.

This is the unsavory story of how mobile phones got made. Amnesty International says that coltan "paid for a war within a war that claimed hundreds of thousands of civilian lives and subjected millions of others to a humanitarian catastrophe." But, despite the UN investigation, nobody has been called to account. And the efforts of the phone makers at the time to clean up their supply chain were, to say the least, half-hearted. Companies like Nokia and Motorola say they first learned about the trade out of Congo in early 2001 and asked their suppliers to avoid purchasing tantalum from there.

But the fact is that for a while the Congo was the main source of the world's coltan, and much of the world's coltan went into mobile phones. Go figure. I'd be very surprised if there wasn't Congolese tantalum in Joe's phone.

Coltan prices eased in 2002 and the Congolese civil war subsided soon after. But the warlords are still in business—mining coltan and making trouble in the jungles. The two are undoubtedly linked. And since then another vital resource tied to mobile phones has come under their control. Rising global demand for tin—driven in part by new rules banning lead solder in phone capacitors—has triggered a rush for rich Congolese reserves of cassiterite, an ore containing tin oxide. The same soldiers, and the same illicit supply channels through Rwanda, are involved.

Joe didn't use his phone for long. Most mobiles are discarded within two years—that's almost 100 million every year in the United States alone. The value of an individual metal in an individual phone is probably only a few pennies. That's why most phones end up in the local landfill. But collectively this is crazy. All that copper and silver and gold and tantalum wasted. All that arsenic and antimony and lead and other toxins leaching into the ground. So deciding what to do with Joe's old phone, I pondered the alternatives.

Recycling seemed the obvious bet. The new European directive on electronic waste, charmingly called the WEEE directive, is big on ensuring that the materials get recycled, but much of this seems to be carried out illegally and dangerously in China and India. In any case, why not prolong the life of the phone itself? Find someone else who wants it.

The mobile phone industry has begun take-back schemes. In Britain, Fonebak claimed to have processed 6 million phones by the end of 2006, of which around two-thirds were reused and a third sent for recycling. But I read an independent study that said that many of the recycled phones from such schemes ended up being exported to untraceable companies. I wanted to be certain where Joe's phone went. I suggested to Nokia that they might like to help me track the final destination of a phone sent to them for refurbishment. Helsinki seemed keen, but their London office stopped returning my e-mails. Perhaps they got bored.

Then I spotted a sign in a shop in Sussex called Cookshop, promising to collect old phones and send them for resale in Africa. Their agent, Phones for Africa, turned out to be a small enterprise that collected around sixty phones a month and was run by Paul Joynson-Hicks, a photographer in Dar es Salaam whose uncle is the head of the small Cookshop chain of stores. I already had a trip to East Africa planned. And everyone along the line seemed keen to help me hand on Joe's phone in person to a new owner.

Everything worked like clockwork. Paul was a bit vague about how the phone got to Dar. Import duty is a bit of an issue, I gathered. But two months later, on a hot and dusty Saturday afternoon, I found myself standing at a tiny kiosk in a litter-strewn slum street in inner-city Dar, briefly reacquainted with Joe's phone. Eventually, my buyer showed up. And that was when I had a surprise. Ally got out his cash and handed over 40,000 Tanzanian shillings. That was the equivalent of $34. But, two years before, Joe had bought the phone new for only $20. Paul smiled when I asked about this later. "That's the going rate here," he said. "Actually I marked the price down a bit to make sure you got a sale." He said the locals like slightly bashed reconditioned phones because they know they are not counterfeit—unlike some on sale in shops down the street.

I didn't quibble. The deal was done. Ally, a student of similar age to Joe, was unaware of how market forces had played out in his case. But he said the phone was cheap. And he was happy to be the second of four adult brothers to get a phone. His father, a preacher, had given him the money. Then he headed off for evening prayers at the local mosque.

So my buyer was happy. And, in the end, my qualms about the price were mollified by the discovery of what happened to the money raised by the sale. The next night, Paul took me to a workshop behind his studio on the city's outskirts. It was dark because the power was out. In the gloom, he called Justin over. It took a while. Justin had no legs, only a couple of stumps. But he did have a blowtorch. Justin, it turned out, was one of a team of twenty-five polio victims that Paul had recruited over a couple of years, mostly from a nearby traffic island favored by beggars.

The profits from selling phones like Joe's went into training them as welders, and buying equipment. Paul—a gangly, amiable guy able to win over industrialists as well as beggars—begged scrap metal and surplus stock from local traders, and the team was turning out a constant stream of rather fetching metal sculptures that sold in local art stores, on the Internet, and through direct commissions.

Wildlife sculptures are the big sellers. The Wonder Welders catalogue (check it out at www.wonderwelders.org) includes dragonflies and gazelles, crabs and turtles, rabbits and giraffes, crocodiles and chameleons, antelopes and aardvarks. Once, the welders made a life-size metal rhino to a detailed spec drawn up by a zoologist. More modestly, I brought home a small warthog made of nails, a spring, a couple of links from a bicycle chain, and some copper wire for its bushy tail. It is sitting on my desk as I type this—a strange swap for Joe's phone.

The Wonder Welders are doing such good business that Paul and his workshop manager, Ellie, have diversified. Paul introduced me to disabled women boiling up paper waste to make new special paper for photo albums, wedding invitations, and novelty cards. They sometimes incorporate scraps of metal, too. The next February 14, my wife received a pretty Valentine's Day card featuring two flamingos decked out with the metal from a Castle beer can. The workshop makes soap, too. And glass blowing is next.

In the back streets of Dar this hive of activity was hugely impressive. There is clearly a limited market for Wonder Welders, though I would guess that dozens of African cities could sustain similar enterprises. But the imagination and creativity and energy that have gone into this scheme show what unexpected potential there can be for recycling our discarded gadgets, and how unexpectedly wide the benefits can be. Wonder welders, indeed.

At any rate, I was happy with the outcome of my decision to find a new life for Joe's old phone, and I knew that Joe would have been just as pleased. A student got a reconditioned phone at less than Dar shop prices. Polio victims are getting off the streets and into creative jobs that are the envy of their fully limbed mates. Some of Dar's growing piles of scrap metal get recycled. And I get to smile at that warthog. I flew out of Dar feeling that my footprint was, for once, positive.

Unexpected Heroes   24
The Queen of Trash and Other
Chinese Titans of Recycling

Cheng Shengchan is an ebullient self-made man with curious habits, like rolling his trousers above his knees while making tea for his guests in his office. He runs what has a reasonable claim to be the greenest, most socially inclusive, and "sustainable" paper mill in the world. And the much-derided paper recyclers of Britain can claim a significant part in his success. Here is his checklist of achievement. It is enough to make any European executive in charge of corporate social responsibility green with envy. "My mill runs entirely on wastepaper," he told me as he served tea. "It is fueled entirely on biofuels; its labor force is made up of the disabled; and not a drop of effluent leaves the site."

Cheng took over the loss-making CXXR United Paper Mill, outside the southern Chinese city of Xiamen, when the state decided to shut it down in 1994. He was a manager back then, though he says now that he had no control, no supplies of raw materials, and few markets. But he was confident he could make a go of it, and bought a 50 percent stake. With the state a sleeping partner, he has run it as his own business ever since. And business is good. After he had introduced me to his son, who he hopes will one day take over the plant, we headed into the yard. Truckloads of wastepaper from local factories were arriving constantly. It accumulated so fast that the mounds of waste almost engulfed the staff as they shoveled it onto conveyors for mixing with water to boil up into a pulp.

Cheng said he uses no chlorine or other chemicals. The pulp just gets a dose of palm oil to help it solidify, before being pressed into giant sheets of thick brown paper that are rolled up and sold, to be made into laminated boxes. Meanwhile, the wastewater squeezed from the new, drying paper is constantly recycled. I saw a fair amount of evidence of that. But his settling ponds did create a sludge that must have been cleaned out from time to time. Still, there were no streams of noxious effluent running into local rivers. And there was no ambiguity about how Cheng fueled the factory boiler. Coal was a thing of the past. Now he burned rice husks, bought from a local food factory. Cheng said the husks cost him less than half as much as the coal he used to burn. And since the husks would have otherwise been left to rot and release their carbon into the air, this was a genuine carbon-neutral biofuel.

Cheng was jovial and candid. He employed handicapped staff because it brought tax benefits, he said. "We qualify as a welfare company. It is easy work for them, so there are no problems." The staff seemed to agree. But I wasn't sure about the health and safety ethic. Where were the guardrails and the masks to filter the dust? But this was China and the place seemed fun. One woman worker darted this way and that among the piles of paper to avoid being photographed. Another male worker pulled a handle to drop a big roll of paper at my feet with a loud thump—just to make me jump.

The waste cardboard was coming in from local beverage- and food-packaging plants. But besides this local waste, Cheng said, he bought container loads of European wastepaper imported at Xiamen docks by a big local trading conglomerate called Xiamen C&D. He took several thousand tons each year, most of it from Britain. Xiamen C&D failed to identify the UK source for me. But its local rival, BCEL, was more helpful. It provided me with a list of all the foreign companies with approval to supply China with paper waste for recycling. In a list of several hundred, there were forty-seven U.S. companies; but only ten British companies made the cut. It seems that while U.S. recyclers generally had a clean bill of health, delivering well-sorted waste, British firms were less highly regarded. Ill-sorted paper waste from Britain has been a well-recognized problem in China. And many British firms were excluded when Chinese customs officials tightened up on the flows in 2004 to make sure there was less contamination of the valuable raw materials with general garbage.

Cheng manufactures more than 38,000 tons of paper a year. And his biggest customer is Top Victory Electronics, a Taiwanese company with a factory 125 miles up the coast that makes liquid crystal display screens for TVs. So, in the wonderful world of globalized recycling, the boxes made from paper exported from Britain and the United States to his mill probably come back to Britain and the United States on the next available container from Xiamen harbor.

It is amazing how much waste gets recycled in China. Earlier in the day I had visited an old garage in the heart of Xiamen where local street sweepers and cleaners brought cardboard, cans, old plastic bags, and glass and PET bottles scavenged from their daily rounds to a man called Mr. Fu. He paid surprisingly good money: about 4 cents for a glass bottle and about 3 cents per aluminum can, which is slightly more than the going rate at Latchford Lock in Britain. Some street cleaners made half their income this way, he said. He got occasional supplies of imported European mixed waste, but it was badly sorted, so there was a lot of material he could not recycle and had to dump. Same old story.

Fu made a good profit. As I watched a rat scuttle away from behind a big cardboard box that once contained a Sanyo TV, Fu said he bought used PET bottles for 4,000 RMB (over $500) a ton and sold them on at 7,000 RMB a ton. That is roughly half the price that the local Coca-Cola plant paid for new PET, I later discovered. His turnover was 3,000 RMB (about $400) a day. He said there were five hundred other traders like him spread around a city of maybe a couple of million inhabitants. It sounded as if remarkably little recyclable material escapes their freelance collectors.

After seeing Fu, I visited a large junkyard entirely devoted to plastic waste: drums and bottles, toys and flip-flops, baskets and hoses, bits of cars and old phones. Mr. Wong said he sent out his truck to people like Fu, and brought back more than 3 tons of plastic a day, for which he paid maybe 15,000 RMB. He seemed to have more than he could cope with. There was about three months' stock in the yard ready to be cut up and shredded and sent off for reuse somewhere. But business was good. The nice BMW parked in a corner of the yard was his.

All this was technologically primitive, but evidence of a well-developed industry for sorting, handling, and ultimately recycling a wide range of materials that are unlikely to find a home in Europe other than a landfill. And it flourished wherever there was waste—in Chinese cities, or where foreign container loads docked. Wong said imports of waste plastic didn't reach him because they were snapped up by people nearer the port in Xiamen. But elsewhere I heard of vast amounts of plastic waste being imported to Hong Kong and shipped on barges up the Pearl River delta to recyclers around the industrial cities of Dongguan and Shenzhen. There, whole villages specialize in specific types of plastic waste, like PET bottles or plastic bags made of LDPE (low-density polyethylene).

The trade in trash sounds bizarre, but from the Chinese end makes eminent economic sense. With so many containers going back to China empty, it costs only about $200 to hire a 40-footer to ship waste for recycling from London to Shanghai, which is a tenth of the price of the journey the other way. Where the problem lies is that it is cheaper than the landfill charges in Britain. So the temptation is to cut corners and send unsorted rubbish. But make no mistake, this is big business.

So yes, we are sometimes sending rubbish to China. But it beats me why people imagine the Chinese would buy our rubbish in order to put it into a landfill. The truth is that they buy it because, in a country with a desperate shortage of raw materials of all sorts, anything that can be recycled is valuable and will be sorted and turned into new products. China needs and uses most of the estimated 2.2 million tons of "rubbish" exported there from Britain each year. If there is waste that does get dumped, it is because British suppliers are falling down on the job of sorting the waste properly prior to dispatch. Once in China, paper and cardboard become the packaging around more products heading back to Europe or North America on the next container. Plastic will often be made into low-grade products like plastic drums and chairs and toys, or new bags. PET and glass bottles get turned right back into new PET and glass bottles.

China has a "queen of trash." Her name is Cheung Yan, and hers is an amazing story, now quite famous in China. Cheung's parents worked in textile factories, and after her father was jailed in the Cultural Revolution she signed up for the sweatshops as well, supporting her mother and seven siblings. Then she got a job in a small paper mill and worked her way up, opening her own recycling business in Hong Kong in 1985.

A decade on, she spotted a chance to expand when China banned logging after big floods in the Yangtze River in 1998, and faced a chronic paper shortage as a result. She and her husband went to the United States and drove round in an old minivan, begging garbage dumps to sign deals for shipping their wastepaper to China. It worked like a dream. And today that business, called Nine Dragons, has become probably the largest paper-recycling business on the planet, buying more than 6.5 million tons of wastepaper a year, mostly shipped by container from Los Angeles and Rotterdam.

The imported waste goes direct from Chinese ports aboard a fleet of five hundred trucks to her huge mills at Dongguan in the south and Taicang, near Suzhou on the Yangtze delta. Each is close to the main exporting zones for Chinese manufactured goods. Millions of the "Made in China" boxes being shipped to Europe and North America these days are made of corrugated cardboard that comes out of Cheung's giant paper-recycling plants. In late 2006, Cheung was declared mainland China's richest person, man or woman, and is probably the richest self-made woman in the world. By the time this book is out she hopes to have surged ahead of paper giants like Weyerhaeuser and Smurfit to become the world's biggest manufacturer of packaging.

Nine Dragons is leading a huge global surge within the paper industry toward recycling. The industry is close to a milestone where more than half of all the paper and cardboard produced worldwide comes from recycled material. Chinese entrepreneurs, it could reasonably be said, are leading the way to a future in which the "loop is closed" on major materials flows, and recycling becomes the norm. There may be doubts about the energetics of paper recycling in Britain, compared to incineration for power, but in most countries there can be little doubt that recycling is the green option for paper waste.

This description of the Chinese recycling industry sounds too good to be true, and of course it is. Not everybody in the recycling business is like Cheung. My swift reality check came just down the road from Cheng's CXXR United Paper Mill outside Xiamen. I dropped in unannounced at Xiamen Yixiang Metal Products, where they recycle aluminum scraps from local can makers to turn them into new ingots of aluminum. To put it mildly, what we saw was a far cry from the way it is done at Latchford Lock. We stumbled on a primeval plant. I watched as men in slippers and without any protective gear worked twelve-hour shifts, dunking lumps of waste metal into cast-iron pots suspended in an open kiln beneath the factory floor. The metal was heated to 1,300°F, at which point the aluminum melted, and then slipper-men scooped up the molten metal in an open, long-handled ladle and walked it across the floor to pour it into open molds.

We walked around for half an hour, unsupervised, watching the men at their dangerous work, sweltering in the heat of the kiln and breathing in the acrid fumes. Then the boss, Mr. Liao, showed up and invited us into his office. He could recycle up to 8.8 tons of aluminum a day here, he said. Since the molds contained 14.8 pounds, that was more than a thousand hand-pourings of molten aluminum a day. Liao used to import foreign cans, but the customs people had stopped it, he said. Nonetheless, he was part of the global aluminum economy. He bought his aluminum at roughly 8 RMB (about $2.20) a pound, and sold it at around 10 RMB, in line with the international price on the London Metals Exchange.

I was about to ask him about the lax health and safety conditions at the plant when he began complaining that regulations in China were too tough and his boss was thinking of moving to the Philippines. And it turned out he was proud of conditions. He brought out of a back room the framed certificate, dated 2005, of his accreditation with ISO 9001, a reward for good documentation of working practices. Our jaws dropped. "That must have cost a good lunch," said my companions later. Sure. This is China.

E-waste

What to Do with That Old Computer

I have a slight phobia about getting rid of old computers. It's not that I've left anything embarrassing on the hard drive; it's just an uncertainty about what to do with them. I left an old Amstrad in a house I emptied before moving fifteen years ago. And I have four computers in my current loft. I presume landfill is not a good idea. About two-thirds of the heavy metals in landfills these days come from electronic waste like computers and mobile phones. At the waste transfer station, they often put computers and monitors to one side. But what then? Sale to Dubai waste brokers for shipment to India, China, and African recycling centers such as Lagos in Nigeria?

Around a million old TVs and computer monitors are exported from Britain to developing countries every year. (There are no precise statistics, but between 165,000 and 385,000 tons of U.S. electronics waste ends up being sent abroad each year.) The precise legality of such exports seems to depend on the declared motive for shipment. Disposal is banned, but reuse, repair, and recycling are generally okay. But of course what it says on the paperwork may be rather different from what actually happens in some distant country when no officials are watching. It was the reality practice that I wanted to know about. And that took me to Mandoli, an ancient Indian village now being consumed by the sprawling metropolis of Delhi. Mandoli is one of the charnel houses of the computing world, where the printed circuit boards in discarded computers end up.

Mandoli was clogged with trucks and oxcarts and dirt. We drove to an area of wasteland, probably a former quarry, known locally as Gadda, meaning "the ditch." That had it pretty well. The ditch stretched for several miles toward the haze of Delhi. A huge sewer pipe from the city ran down it, and not far away was a narrow but persistent stream of nasty black fluid from a fly-by-night operation dying denim jeans for a city garment factory. Then we came across a smoldering pile of black metallic-looking ash with a nasty acrid stench. And farther on was an area with a rudimentary network of alleys and brick walls about head high. It looked as if someone had decided there was no need to finish the construction of buildings. Open-air workshops would do just as well.

The alleys crunched under our feet. They were covered in bits of old yellow-and-green computer circuit board. We opened a gate and entered one of the workshops. "No, there are no circuit boards here," said a youth squatting on one of the half-built walls. When I pointed out a pile of them right by his wall, he shrugged. "We wash them to get the copper," he said. How? "In acid."

And there was ten-year-old Rajesh doing just that. He was dunking circuit boards into an old blue oil drum that contained acid. The drum was almost as tall as him; he was on tiptoe to complete the task. The boards would stay in the hot acid for several hours, he said. During that time the thin layers of laminated copper within the boards would be released.

Rajesh had thick rubber gloves. They were a bit big for him. But so long as he kept them on, they protected his hands. But as he leaned over the edge of the drum, he splashed his thin trousers and torn Adidas T-shirt. He had no goggles or mask, nor were there any lying around. The whole process was perfunctory, hazardous, and wasteful. There was a pool of discarded acid in the dust. Sometimes they poured the spent acid into the open sewer by the road.

Rajesh had come here with his brothers a couple of months before, from the poor Indian state of Bihar. Bihar is notorious for supplying gangs of laborers to work on India's booming construction sites, do the hard labor on farms, and—as here—take on the nasty dangerous work that nobody else, even in India, will do. Rajesh wasn't the youngest person working here. His younger brother, who looked about six years old, darted among the drums behind him. He only seemed to be doing odd jobs. But he was coughing. In the oppressive heat of the day, the acrid yellow-green fumes from the drums were trapped behind the brick walls. I too was coughing after five minutes in the place.

"When new workers come here, they suffer giddiness and headaches from the fumes," admitted the older, shaggy-haired brother of the other two. "But at the end of the day we have a strong drink and we are okay," he laughed. "We Indians are strong. And it is better than having no job. Before this I was unemployed in Bihar." His name was Rakish. "I'm fourteen years old, I think," he said. He looked old before his time.

Next Ajay showed up. He was twenty-five and in charge of the seven-person workshop. He worked directly for the owner, but wouldn't say who that was, only that "he lives in Delhi." Ajay said he went to the local market to buy sacks of waste circuit boards. Usually, they had already been stripped of capacitors and other valuables, so they were cheap. A bag containing more than 2 pounds of boards cost between 5 and 10 rupees, or less than 30 cents. Once dunked in the acid for a few hours and scraped clean, they could yield a pound or more of a red sludge containing copper, which he sold for more than 100 rupees a pound.

Copper prices had quadrupled in just five years. And though he was hardly in a position to drive a hard bargain, Ajay knew it was a seller's market. He sold to traders, but I guessed the eventual market was two works manufacturing copper wire, both within a couple of miles of the village. The leftover plastic boards could sometimes be sold. Otherwise, Ajay said, they just dumped them in a ditch, or set fire to them. Ajay's unit processed between 30 and 45 pounds of circuit boards a day, he said. So that might provide an income of a few hundred rupees. Ajay had to buy the acid, of course. And he paid the workers 50 rupees a day, a bit over a dollar, for the privilege of wrecking their lungs.

It was a meager business. But this unit, which measured about 10 yards square, was one of about twenty clustered around the alleys of Gadda. There were maybe a hundred workers, including other children, all engaged in the same activity. Each had a manager, like Ajay, who worked for an absent owner. The workshops were full of bags containing old circuit boards, stamped with household names from the computer industry. One board I picked up at random was from Invensys, the British-based computer systems company. Intriguingly, some bags contained bits of unused boards, bright and clean but broken into pieces. They appeared to be cast-offs from manufacturing.

I went to Mandoli with Priti Mahesh, a researcher at Toxics Link, an environment group based in Delhi. She mentioned that she had seen boxes of broken pristine circuit boards, all with labels from Akasaka Electronics, based in Mumbai, and from Samsung, for whom Akasaka made circuit boards. She thought some waste-disposal subcontractor had sold the broken stuff in Mandoli.

As we talked, trolleyloads of wire were being unloaded from a truck and divided up among the units. There were piles of wire all around. Someone was sleeping on the stuff; it made a good mattress. Tests on the environment around the workshops found heavy metals in the soil, caused by a mixture of ash from the fires, fallout from the toxic fumes, and residues in the spent acid poured onto the ground. Yet nobody there was concerned. I spotted one vacant plot right next to a copper recovery unit where someone was growing vegetables. The leaves would certainly have been heavily laced with toxins.

Most people in the units were temporary workers, but I met Javed, who had been in the business for ten years. In his blue pullover with a Duke logo, he was better dressed than most. Javed denied the business was dangerous. "Some people get burned when the acid is splashed but we don't get major injuries," he said. "They have rubber boots and gloves and masks." I only ever saw gloves; and some units didn't have them.

The delivery of circuit boards seemed quite organized. Trucks come to the Mandoli village market once or twice a month, Javed said. "In the early days, we had TV and radio circuit boards. The computer boards only started turning up in the market about five years ago." That was interesting. It seemed Mandoli started receiving the boards only after the Indian government passed laws making it illegal to import electrical waste of this sort. Presumably, a black market developed in place of the legal trade—and ten-year-old children got roped in to do the dirty work. Graft was clearly endemic. "The local police come and harass us all the time, so we have to pay them," said Javed. But the public spotlight on places like this was growing. A year before, police shut down a similar operation in Silampur after some media reports. He reckoned big recycling firms would take over eventually.

But for now, these small backyards form a significant part of a much larger industry. Ravi Agarwal, director of Toxics Link, estimates that two-thirds of the e-waste in Delhi's scrap yards is imported. E-waste comes from Southeast Asia, Europe, and the United States, frequently after being traded by brokers from Dubai. That city has a long tradition of smuggling goods to India, and ways and means of getting past customs controls at Indian ports. A generation ago it was gold. Now the government bans imports of e-waste, and the Dubai traders are neck-high in that business.

The containers arrive at ports on the west coast, such as Mumbai and Kandla. An estimated one thousand truckloads of electronic waste come up the crowded highway from Mumbai to Okhla in the south of Delhi each year. Scrap dealers in Okhla auction off the waste, which may turn up later in Muslim neighborhoods on the outskirts of Delhi, such as Silampur and Mayapuri and Turkmangate, that specialize in dismantling computers and other e-waste. Perhaps a thousand PCs make this journey every day.

In Silampur, marketers who once traded in fruit and vegetables and rice now set to work. As we toured the district, it was often hard to tell if they were computer repair shops or recyclers. Monitors and their casings, the chips and capacitors and gold-plated pins and solder from the circuit boards, the stripped motherboards, the keyboards, disk drives, mice, and other components all have their disposal routes, and their own specialists adept in the dark arts of low-tech, high-risk materials recovery, with hammers and screwdrivers and pliers and acid.

Circuit boards alone may go to several specialists. One to take the chips, another to remove gold and silver, another to melt off the solder, and, finally, the copper collector. Mandoli is pretty much the end of the line. Cables and transistors are stripped of their copper by burning and boiling. In Shastri Park in east Delhi, they smash up old monitors and dismantle printer cartridges, creating clouds of toner dust, some of which is collected and sold on. Plastic PC casing often ends up in another Delhi suburb, Mundaka, which houses Asia's largest market for waste plastic.

Delhi is not the only place where these activities are carried out. Sher Shah, a suburb of Karachi, Pakistan's main port, is another. One study in Lagos found five hundred containers of e-waste, with perhaps 100,000 old PCs inside, arriving at the port each month. A cluster of villages known collectively as Guiyu, near Santou in southern China, has become notorious partly because such activities are much more visible in this formerly rural region than when spread across a large urban area like Delhi.

In high-wage economies, e-waste has always been regarded as a liability to be disposed of as cheaply as possible. But in low-wage economies, where people will work for virtually nothing and in poor conditions, e-waste is a resource. It can be profitably mined for tiny amounts of valuable raw materials. This is recycling: closing the loop. We should all be in favor. But in places like Mandoli, the human cost is too high. The air is rich in mercury and dioxins; the drinking water is tainted with toxic metals. The children are scarred or worse by acid, their lungs blackened and diseased by fumes; and an air of exploitation and graft permeates the wider society.

How responsible should we feel for what happens here? In 2007, Britain enacted new European laws, the WEEE directive, to increase recycling of computers and other electronic waste. It requires the producers of waste to find a safe disposal route. The aim is to reduce landfilling, but the risk is that it will result in a surge in exports of e-waste to places like Mandoli. Britain alone produces more than a million tons of electronic waste a year. That's the weight of 2,400 jumbo jets—and includes some 55,000 tons of scrap circuit boards. Half a million old and warranty-returned TVs, 100,000 video recorders, and 120,000 computers are exported. Brokers pay $30 for old TVs, whether or not they are working.

There are already hundreds of e-waste traders exporting British e-waste. Many of them are not based in Britain, and often do not even declare their addresses, say industry insiders. You reach them by mobile phone or via websites. A report for the Environment Agency says they "do not seem keen to divulge information about the destination or fate of equipment and will only discuss the equipment they want and what they are prepared to pay." Cases in the United States involving the shipping of electronic trash to Hong Kong suggest that the trade operates in a similar twilight world between legal and illegal.

And now there is the WEEE directive, which requires that "recovery" rates for e-waste—that is the proportion not going to landfill—have to rise from 15 percent to at least 75 percent. But where will all this end up? In special facilities? Or in the Dickensian workshops of Delhi or Lagos? In theory, the retailers and manufacturers are responsible for what happens to their products. They have to take back old products and find them a safe disposal route. The paperwork will all be checked and the ultimate fate of the material ensured.

Really? How scrupulous can we expect them to be? Who is physically going to open the containers destined for Singapore or Mumbai or Dubai or Shenzhen? Who is going to judge whether the contents are capable of being safely recycled and sold in a distant land, where labor is cheap and pockets empty? Untangling the paperwork will be hard enough, let alone deciding if any of it is truthful. The WEEE directive looks increasingly like a bogus recycler's charter. Certainly in India they expect that the new directive will only increase the number of shipments to India. That will be good if the recycling is genuine and safe. Bad if it all ends up in Mandoli or Guiyu.

I went in search of some way I could dispose of my computer with a clear conscience. I went first to what is advertised as "Britain's largest electronics recycling plant." Owned by the Wincanton trucking group, it promises to handle up to 11 tons of e-waste an hour—more than 82,000 tons a year. Millions of consumers and many retailers will be entrusting their e-waste to this facility.

Standing in the warehouse, I saw a sample of the wares already coming to Billingham. There were bagless vacuum cleaners, Compaq computer mice, Game Boy control consoles, miscellaneous keyboards, and old phones. On other days it could be washing machines and vending machines, lawn mowers and TVs, medical equipment and electric toothbrushes, toasters and vacuum cleaners. Much of it was unsold stock from shops, in unopened boxes, and sometimes entire pallets of single items. There were piles of boxed-up Bush VCRs, unsold even when knocked down to $49.95. This was stuff going for recycling before it had been used once. What a waste.

A conveyor belt of computers disappeared into a large hopper. Inside, two heavy chains whipped around at high speed and broke up the equipment. Out came mangled, fist-size piece of e-waste. A dozen or so staff in overalls picked through these gobbets of our cyber-economy and put them in the right bins, marked for batteries, cables, PC boards, copper and iron components, stainless steel, aluminum, and so on.

I had expected to see real recycling going on. Instead it seemed to be the British equivalent of Silampur, where they dismantle the e-waste and pass it on. "Basically we just separate the toxic stuff from the recyclable stuff," said general manager Chris Wilson. "What we do could be done by men with screwdrivers. But we'd need about two hundred people, rather than ten people." The only real difference is that in India it is cost-effective to use people; in Billingham chains work out cheaper.

Out at the back the different waste streams were assembled for dispatch. A lot of metals go to Liverpool-based European Metal Recycling (EMR), the biggest scrap-metal merchants in Europe. EMR in turn sells it on to Corus, formerly British Steel, or to steelworks abroad. Most circuit boards, with their chips intact, go to high-tech metal-recovery plants such as the Boliden refinery on the edge of the Arctic Circle in northern Sweden. But many recyclers cherry-pick the boards with the most valuable metals, like gold, silver, and lead. They don't want the rest, which is why they go to India and China right now.

Batteries are recycled in Britain. Cathode-ray tubes used to be recycled to make new tubes. But now everyone wants plasma and LCD screens, so that route is gone. There is a growing cathode-ray-tube mountain. Old copper cable goes through a high-tech version of the Mandoli operation, with PVC stripped off prior to making new cable. Plastics go to China.

That's the theory. But in practice the waste goes through a chain of brokers. And already much e-waste goes east. So what happens when the volumes increase fivefold? I was left more worried than reassured by how the directive was being implemented. Chris thought it was the Environment Agency's job to regulate the brokers. But at the agency they say the onus is on the exporter. So who were the brokers? I asked. Maybe I could check them out myself. "We use a number of brokers," said Chris. "We just look for the best deal on the day. We won't name them because of commercial confidentiality."

A big problem is that the WEEE directive is all about the environment and does not have any specific health and safety requirements. Recycling is recycling is recycling, whether at high-tech Boliden or low-tech Mandoli. When licensing brokers, the Environment Agency can't insist on socioeconomic conditions. Chris couldn't guarantee that his broken bits of old computer motherboards wouldn't end up in Mandoli. His only thought was that Billingham's relatively "clean" waste streams could usually command high prices, which made the dodgier waste practices unlikely. I wasn't sure of the logic of that. It might prevent the majority of the waste from being dumped. But that doesn't happen much anyway. And an Indian copper-smelting company would still buy a container load of old circuit boards, and still send them down to Mandoli for the Rajesh treatment, if that was the cheapest option.

We probably have to accept that micromanaging recycling in distant countries is not possible. And perhaps not desirable either. Countries like China and India need raw materials to maintain their fast-growing industrial sectors. Increasingly, that is coming from the recycling of products previously sold to the rich nations. As Chris said, "We cannot say for sure that WEEE won't stimulate poor recycling conditions in India and other places. It may. But at the end of the day, if they don't recycle, then that would stimulate more primary-materials production—more mining and refining of ores—which would have its own health and safety downside, its own social problems, and its own victims."

So what should I do with my old computers? My first inclination is to leave them sitting in the loft. They may not be giving up their precious metals to anyone for recycling, but at least they are doing no harm either. But surely there must be a better way. Finally, I found it.

A couple of months after visiting Delhi I went to a workshop in Nairobi where they take in container loads of old computers from Europe and North America. But rather than breaking them up, they refurbish them and ship them out to secondary schools. The scheme is the brainchild of social entrepreneur Tom Musili, who had seen something similar going on in Canada and thought he could bring it to Africa.

Tom began refurbishing computers from a bench in a spare room at a local school in 2003. Computers for Schools Kenya (CFSK) now has a small staff and is busy setting up a network of regional centers to handle the increasing numbers of old computers it is being offered, both from Kenyan corporations and from around the world.

Container loads of computers arrive regularly at Mombasa port on the coast and come to Nairobi by train. The day I was there, Tom's live-wire coordinator, Norman Makinga, had received a consignment from the University of Bradford in Britain and was tracking the progress of another container load from Britain. "We get an average of one container, that's about four hundred and fifty machines, each month. So we are handling about twenty machines a day." The operation is straightforward and efficient. Incoming machines are coded, cleaned, and checked over. About a fifth of them don't work or don't meet the specs. They are cannibalized for parts for the rest. With the rest, the memory is wiped and new software installed. A deal with Microsoft means there is no charge.

Norman's team packs up computers in batches of twenty, ensuring one machine between two pupils in classes that typically have forty children. Before Tom came along, most schools in Kenya, even large secondary schools, had no computers. By late 2006, Tom had supplied some 300 Kenyan schools with more than seven thousand computers, and there were another 850 on the waiting list.

There isn't much storage space in the workshop, so typically computers are on their way to schools within four days. I looked at the schedule pinned up on the wall. It was a rare day when some school somewhere in Kenya did not get a new computer lab. Computers go to the most remote areas of the country, where the constraint is not broadband access (forget it; there isn't any) but electricity. At the end of my visit, Norman was just shoe-horning one last keyboard into an old Ford van with forty-five computers destined for three schools in Garissa, a poor, dusty, rebellious town close to the border with Somalia.

A small party gathered to see off the driver on a four-day journey. The roads out there are bad and the equipment could get damaged, so Norman had taken the sensitive hard drives out of the machines and protected them in bubble wrap. "We'll put them on the front seat," he said. It was the last leg in the long journey those computers had made from Bradford in England.

What is the lifetime for these refurbished computers? "We reckon that a decent branded computer has a lifetime of around ten years," said Norman. "So we won't take computers more than six years old—Pentium III 550 or better, please. Sorry, but they don't take Apples." Every day, four or five machines don't make the grade. Where do they go? Norman took me to a second workshop. "We recycle virtually everything here. But we work hard to make sure there is no bad practice." In the yard there were piles of mice and keyboards and casings, both metal and plastic, as well as motherboards and wires and hard drives—even boxes of screws. I got him to take me in detail through the story of what happens to those piles. If I was going to give one of my computers to CFSK, I didn't want its motherboard to end up dunked into an acid tub by a ten-year-old street kid.

There is a market for almost everything, he said. Hard disks sell for 30 shillings each (about 40 cents), aluminum casings for less than 25 shillings a pound, and copper wire for 60 shillings. Metal casings get turned into guttering. Aluminum is recycled in Nairobi, but hard plastic goes to China. "We even take the metal balls out of the mice and they go for recycling too, mainly for their lead." He handed batteries over to recyclers, "but we have to keep a close check, because some of them will just pour the acid they don't need into ditches." Hmm.

Upstairs, he had a big collection of monitors stored up. Some he would sell to a college for students to cannibalize them. "We sometimes let artisans have the monitors to build cheap TVs round them, and sell them in the local Jua Kali market," said Norman. "But the circuit boards are the most difficult thing." He has a buyer, but he wanted a container load at a time. That would take six years and Norman wasn't sure where they would end up even then. Delhi, maybe?

Norman clearly works hard to find a safe home for the detritus from his computer refurbishments. It is not perfect. I would imagine that things do fall into the wrong hands. But is this a good, valuable business—a worthy end for my computer? I think so.

Around half of CFSK's computers come from Britain via a London charity call Computer Aid International. CAI is the world's largest not-for-profit provider of computers for developing countries. Its donors range from British Airways and Mitsubishi to Christian Aid and a bloke who lives down the road from me. Its boss and founder, Tony Roberts, has fought continued skepticism in the environmental community about the morality of shipping old computers out to developing countries.

Some say the risks of them ending up somewhere like Mandoli are too high. But he says we should see the potential good that can be done as well. "We at CAI have delivered seventy-eight thousand machines round the world so far. We believe they have all found good homes," Tony told me in late 2006. He was aiming to hit 100,000 deliveries by CAI's tenth anniversary in October 2007. Three-quarters so far have ended up in Africa, where Kenya takes the most. When we spoke, he was trying to set up a new computer lab in Kibera, the huge slum on the outskirts of Nairobi.

The potential of CFSK to take the world's computers may be limited. But for Tony the sky is the limit. He believes there is huge potential to set up similar enterprises round the world. Most discarded computers have several years of active life left in them. And now that more and more computer owners are having to think about where their old equipment ends up, the potential is growing fast.

Tony had offered to pick up one of my old computers and rush it out to Nairobi so I could see it being refurbished by Norman's team. It was a nice idea, but they would have had to put it on a plane rather than in one of the regular container shipments. Could I really justify those air miles? I thought not and turned that idea down. Anyway, I wanted to see the operation first. Now that I have, I am convinced. I can't think of a better place to send my computer.

Part Seven

My Species and Saving the Planet

Good News from Africa 26
Why We Can Feed the World

It must be true. We've been told it so many times. The overfarmed and overgrazed soils of Africa, especially on the fringes of the Sahara, are losing their fertility and eroding away. As the population grows, poor farmers are mining the last goodness from their land. Their animals graze the grasslands away to nothing and the desert sands move in. Environmentalists say it; development economists say it; politicians say it; soil scientists say it.

"An area the size of Somalia has become desert over the past 50 years. The same fate now threatens more than one-third of the African continent," reports the UN Food and Agriculture Organization. "The main cause is mismanagement of the land." Its sister body, the UN Environment Programme (UNEP), claims that 900 million Africans face starvation as their soils crumble away. UNEP masterminded a UN Desertification Convention in 1996 in an effort to reverse the trend.

But out in the shimmering heat of arid Africa, where tens of millions of farmers scratch a living from the soil, new research suggests that this apocalyptic vision is little more than a mirage. Farmers are finding ways to grow more without destroying their soils. Farm yields are often up, not down. Soils are often getting better, not worse. Fast-growing populations continue to be fed. In places, the desert sands are even retreating. Indeed, for most places at most times, the whole notion of desertification makes no sense.

This rosy picture is not always true, of course. Droughts can still be devastating. And the case of Darfur, the blighted western province of Sudan, is evidence of the combined effects of dry weather, fragile soils, and civil war. But consider, on the other side of the equation, the dusty desert margins of northern Nigeria around the ancient caravan city of Kano. Here, population density has soared to levels similar to Belgium, and some 85 percent of the land is now cultivated. Rainfall is declining, the availability of chemical fertilizer has fallen by 80 percent, and only the richest farmers can afford high-yielding grain varieties or irrigation. The poor make do with small scraps of sandy soils. Surely these fields should be turning to dust as yields plummet, hunger spreads, and refugees head for the cities?

But that's not what I saw when agricultural scientist B.B. Singh, who heads the Kano office of the International Institute of Tropical Agriculture, drove me through the area. The dusty roadsides between the closely spaced villages were busy with fruit and vegetable stalls and behind them the fields were already green with bushes laden with the first black-eyed peas of summer.

We visited Ado, a farmer who tended a 5-acre plot on the outskirts of Badume village, about 30 miles northwest of Kano. Ado was exultant. The previous year, he had harvested just two bags of peas from his plot. This year, he got seven bags for the same effort. He had a good sorghum harvest too. Ado took me behind the high mud walls of his compound to an inner sanctum where the reasons for his new-found success were bleating. He used to let his sheep roam free. Now he had half a dozen tethered here, munching away at straw left over from his fields and creating a large pile of manure to fertilize the next crop.

Sheep manure is transforming Ado's life. "Now I can send my three children to school," he said. "The boys will become farmers, but I want my daughter to become a doctor." His neighbor Galadima was doing the same thing on his 15 acres. "Crops grow much better with manure," he told me. "I don't use chemical fertilizer at all now." His two wives and eighteen children came running out of the house and lined up for a family photo. They all looked well fed.

Singh's researchers confirm Ado's interpretation. But there's another reason for Ado's success, he said. The black-eyed peas are leguminous crops that fix nitrogen from the air and deposit it in the soil. It is the manure plus the legumes that have made Ado's soils so productive. And the same formula is common in the region around Kano, says Frances Harris, a soil scientist at Kingston University in

Surrey. "The zone is supporting intensive cultivation without suffering from land degradation. The key is the integration of crops and livestock, because it enhances nutrient cycling." Legumes and manure put back what the grain crops take out.

As a result, the Kano region, though arid, is the most agriculturally productive part of Nigeria. Yields of sorghum, millet, cowpeas, and groundnuts are increasing despite reduced rainfall. All this in a region where many experts believe only expensive irrigation systems can produce worthwhile crops. Back in his small office in a back street of Kano, Singh is adamantly optimistic. "Even less than twelve inches of rain is enough for good crops. We can double yields here easily and improve the environment at the same time. And we can do it all over Africa."

Harris makes another point—far from being a liability, the high local population densities around Kano are essential to this form of intensive rain-fed farming. In a land where tractors are rare, people provide the labor to tend fields, feed animals, and spread manure. The old environmental shibboleth that rising populations trigger soil abuse and desertification is being turned on its head here. As is the idea that livestock are an environmental curse. Far from it. More people and more livestock mean the land can be looked after better.

All this would be a mere curiosity if the Kano story was a one-off. But similar stories are emerging from all along the Sahara's edge, from Niger and Senegal, Burkina Faso and Kenya. A region that a generation ago was being written off as doomed to desertification is recovering. David Niemeijer, an environmental geographer from Wageningen University in the Netherlands, has spent more than a decade studying the soils of eastern Burkina Faso. "I went there expecting to find widespread land degradation, especially in the most densely populated areas," he says. After all, the area had seen a decline in rainfall and a tripling of population over forty years. "But there was no evidence of land degradation connected to human activities. Nor was there any decline in farm productivity. In fact, yields of many crops had risen sharply." When he compared soil fertility today with data collected during a French survey in the late 1960s, he found no evidence of decline. No surprise, surely, when Burkina Faso produced 36 percent more food at the end of the 1990s than at the start.

So, what are the farmers doing right? They are erecting low walls of stone and earth to keep soil from washing off sloping land in the occasional heavy downpours. They are spending more time weeding and thinning crops. They are adding manure. And they are managing the land better, forming gangs to tend each others' fields during busy times, and lending and borrowing land, livestock, and farm equipment. "Of course life remains hard," says Niemeijer. "These farmers are still never sure if they can feed their family, and they are not always in control of their destinies. But things are not going down the drain."

Caution is clearly warranted. In 2001, I interviewed Boubacar Yamba of the University of Abdou Moumouni in Niger. He told me a similar story of revival from the Maradi district in southern Niger. Desertification had gone into reverse, he said. Well, maybe so. But in 2005, Maradi district was at the center of famine in Niger. The desert margins remain perilous places; but they always have been. The question is whether they are doomed to deteriorate, and the evidence is that they are not. Perhaps the best-researched example comes from Kenya. Sixty years ago, British colonial scientists wrote off the eroding, treeless hillsides of the drought-prone Machakos district of Kenya, east of Nairobi. Soil inspector Colin Mather called the bare hills "an appalling example" of environmental degradation. The local Akamba people, he said, "are rapidly drifting to a state of hopeless and miserable poverty and their land to a parched desert of rocks, stones and sand." Similar reports came in the 1950s.

There have certainly been bad times. But, even though Machakos's population has risen fivefold since the 1930s, most years these hills are greener, less eroded, and far more productive today than before. Farm output per acre is ten times what it was in the 1930s and five times what it was in the 1960s. There are more trees. And with tens of thousands of miles of terraces dug on steep hillsides, erosion rates are probably at an all-time low.

According to Michael Mortimore, a British geographer who has written extensively about Machakos, the Akamba responded to the environmental crisis half a century ago by switching from herding cattle to settled farming. This gave them the incentive—and their rising population gave them the labor—to work the land properly, digging terraces, controlling weeds, and collecting water in ponds for irrigation.

Every farmer seems to have his or her own story of innovation. On the road out of Machakos town, I dropped in on Jane Ngei, who had used an ox plow, spade, and wheelbarrow to dig her own small dam to catch the rainwater that ran down the road past her farm. With bucket and perforated hose, she uses the water to irrigate about 5 acres of vegetables, maize, and fruit trees. The Akamba farmers are reaping the rewards of improvements like these, selling vegetables and milk to Nairobi, mangoes and oranges to the Middle East, and avocados to France. The people of this area are also among those smallholders who are growing green beans for air-freighting to Britain, with no evident damage to their environment. Rather the contrary. The income is encouraging them to keep up maintenance of their terraces and irrigation channels.

Machakos is just a couple of hours from Nairobi, home of the UN Environment Programme, which still maps much of the district as on the verge of desertification. Yet when I asked, they said they had never sent anyone to research the reality. Other experts are starting to question the sweeping claims of environmental decline still being made by UNEP and other UN agencies.

Camilla Toulmin, head of the International Institute for Environment and Development in London, and a leading drafter of the Desertification Convention a decade ago, now admits that "evidence for soil fertility decline stems from a few highly influential studies of land degradation in Africa, which have been quoted over and over again." Much of the desertification story, she says, doesn't stand up to investigation on the ground.

Researchers tend to visit the desert margins so rarely that they have little history to back up their guesses about environmental processes, says Niemeijer. "Every piece of degraded land has been seen as evidence of destructive human activity. But when we asked farmers about degraded land near their villages, they generally said it had been there a long time and was a natural feature."

Deserts did advance in Africa during the droughts of the 1970s and 1980s. But if bad farming was to blame and the processes of desertification were unstoppable, why does the Sahara retreat during years of higher rainfall? Mortimore says the image of desertification caused by human activity has become an "institutional fact," too important for careers and reputations to be lightly dropped.

The truth is not that farmers never destroy soils, nor that deserts never advance, but that there is rarely anything inevitable or irreversible about the process—and that horror stories may misrepresent the reality. Even when soils have been in decline, as in Machakos, farmers have shown themselves capable of turning the tide. Moreover, it seems that even rapid growth of population can be a spur to change, rather than a curse.

The arid lands are only part of the story, of course. Even if poor people living in marginal environments can feed themselves, and maybe sell a few beans to Britain, it doesn't mean that the world as a whole can feed itself. Especially when half the world lives in cities. But there is good news here, too. Over the past four decades, worldwide food production has more than kept pace with the doubling of world population. There is currently an average of 2,800 calories of food available each day for every human on the planet—a quarter more per head than half a century ago. There is enough to feed everyone.

Moreover, there is potential slack in the system. It takes 2 pounds of grain to produce 1 pound of chicken, 4 pounds of pork, or 7 pounds of beef. So if only a third of the cereals fed to livestock were put instead onto human plates, the per-capita calories available daily would rise to 3,000. That is unlikely to happen; meat eating is growing fastest in developing countries. But it nonetheless shows the potential in the world's food supply system.

There are huge problems. The rich countries, with a quarter of the world's population, still take almost half of the world's agricultural products, largely because they convert more crops to meat. In Africa overall, agricultural productivity has actually gone down since the 1960s, while the population has continued to rise. Key crops like cassava and bananas are suffering ever more from diseases. The African bush is emptying of bushmeat, and the world's oceans are emptying of fish, both because of overexploitation. A billion people are malnourished. The problem is now contributing to at least a third of child deaths around the world.

But all these problems have as much to do with the world trade system and bad government as with absolute shortages. Even when outright famines occur, there are usually stockpiles of food close to the starving. The problem has been that the starving are too poor to buy it or are cut off from it by conflict or persecution. In some parts of the world, the critical problem is not a shortage of land but of labor—in parts of AIDS-torn Africa there is nobody to tend the land, and good fields are returning to bush. Or a shortage of water. But even here, there are often cheap technical solutions at hand that could allow farmers (the biggest users of water in most countries) to grow twice and sometimes four times as much food with the same amount of water in many places. The problems are not technical, and often not even economic. They are political and social—how to ensure that what needs to be done gets done. And to ensure that, where necessary, growing food has priority over growing resource-hungry but lucrative export and nonfood crops like cotton and—potentially the biggest threat of all—biofuels.

But one thing gives me grounds for optimism. The sheer ingenuity of people. Farmers, in my experience, are among the most resourceful and inventive people in the world. I can see it on the fringes of the Sahara, where local farmers seem to know much more than the doom mongers of the international community. And I see it too in the cities of the world. For people living in cities are growing huge amounts of food. By some estimates one meal in every five served around the world is grown within city limits. I've begun to believe that, far from urbanization being a threat to food production, it is a spur to it. Large local populations, easily accessed at every roadside, mean that urban farmers have ready markets among people with the cash to buy. And they respond by meeting that demand.

In Kolkata, India, I have seen old waste dumps where twenty thousand people farm the rich compost left behind by the city's garbage, and ponds full of sewage effluent where carp are raised in profusion. In Havana, there are market gardens growing organic produce on every patch of wasteland. In Lima, they raise guinea pigs for meat in squatter settlements. In Nairobi, chickens fatten in coops bolted to apartment walls. In Port-au-Prince, Haiti, people grow vegetables in old truck tires, baskets, and even kettles. In Moscow, scientists grow vegetables in their laboratory grounds as they wait for the paycheck that never comes. Sheep graze the shoulders of roads in the Armenian capital Yerevan.

An estimated 800 million urbanites—a quarter of the urban population of the planet—spend some time each week tending to food plants. In some cities, farming is the biggest industry. Most of the poultry, pork, and vegetables eaten in cities are grown within the city limits. In Bangkok, 60 percent of the land is devoted to farming. In Greater London the figure is 8 percent. In Bogotá, farming provides twenty jobs per acre—more than a supermarket. Cairo has eighty thousand head of livestock. Hong Kong produces two-thirds of its own poultry, a sixth of its pigs, and half its vegetables. In Shanghai, about a third of the land within the city is used for agriculture, and almost a million inhabitants still work on the land. The city produces virtually all its own milk and eggs and most of its own vegetables, as well as much of its meat and 2.2 million tons of grain a year.

In Europe, urban agriculture is synonymous with allotments, where aging city dwellers grow their potatoes and onions on a strip of public land between the railway line and the gas works. But today there are long waiting lists to acquire a plot. Allotments are tended by over a million people in Britain alone, and many of them, contrary to the stereotype, are young. For the rich world, growing your own is in part a luxury and a welcome change from supermarket shopping. For the poor in shantytowns, or in cities where services are breaking down, it is often done out of necessity. For others it is simply a good commercial opportunity. High demand means that urban farmers can make bigger profits than their country cousins. They often invest in novel systems for maximizing yields on tiny plots. Hydroponics, a system in which roots are suspended in a liquid growing medium, is popular in places as diverse as Singapore, Montreal, and Bogotá. Because there is no need for soil, hydroponics is also great for rooftop gardens.

"Urban farming is often minimized as being merely 'kitchen gardening,' or marginalized as a leftover of rural habits. But it goes far beyond gardening," says Jac Smit, president of the Urban Agriculture Network run by the UN Development Programme. And it is, he says, an efficient use of resources. Typically, intensive production of vegetables in urban areas uses less than a fifth as much irrigation water and one-sixth as much land as mechanized rural cultivation.

City farmers even have free fertilizer in the form of human sewage. The distribution of night soil by bucket—once a fixture of urban life from Paris and London to Beijing—is rare now. However, some 10 percent of the world's irrigated food crops are simultaneously watered and fertilized by the smelly stuff coming out of city sewer pipes.

Such methods can be dangerous, of course. Sewage transports disease-carrying pathogens. Leafy vegetables soak up air pollution, especially from car exhausts. Children may play in the back garden, where pesticides have been sprayed. In Peru in 1992, sewage water used to irrigate urban crops helped spread cholera. And close contact with livestock such as pigs or poultry can bring diseases, including Avian flu. For a long time city authorities discouraged or even banned urban farming. But Smit says that view is changing. Recent studies in the Philippines and elsewhere have linked better child nutrition to the local production of food in urban areas. So a better approach would be to help make it safe by educating farmers about the risks, and treating sewage to remove pathogens but leaving the nutrients, for instance.

But my point is this. Urban farming shows how people are resourceful and ingenious. In most places, and most of the time, we can and will feed the world. As B.B. Singh put it to me as we toured the farms of northern Nigeria, "There is no reason why even Africa cannot feed itself."

Beyond the Clockwork Orange

27

Why We Can Green Our Cities

A hundred years ago, the largest city in the world was London, with a population of 6.5 million. When I was born in 1951, the top dog was the world's first megacity, New York, which had just topped 10 million people. Today there are at least twenty cities above that figure, including three each in India and China. Each of these megacities contains more people than the entire population of the planet at the end of the last ice age. The new world urban leader, Tokyo, has mushroomed to 34 million people. Most people now live in cities. And most of the world's population growth is now happening in cities. There are 50 million more urbanites every year, one hundred every minute.

Cities are the economic powerhouses of the world, consuming three-quarters of our resources and expelling their half-digested remains in clouds of greenhouse gases, billions of tons of solid waste and rivers of toxic effluent. Their environmental footprint is catastrophic. Returning the world's population to the countryside isn't an option. So if we are to protect what is left of nature, halt climate change, maintain the planet's life-support systems, and improve life for the world's poor, we need a new form of city living. If the world is to save itself, the journey must start in the cities. Fortunately, the potential for green cities is huge. For cities are also the great innovators, the great investors, and the great drivers of change. Better still, the sheer size and density of cities makes key tasks like recycling and banishing the car easier. Far from being polluting parasites, cities could hold the key to sustainable living.

There are no new eco-cities yet. But many cities have showcased eco-projects. For example, at the new $40 million home of Melbourne's city council in Australia, hanging gardens and water fountains cool the air, wind turbines and solar cells generate most of the electricity, and rooftop rainwater collectors supply most of its water. And Valencia on Spain's concreted Mediterranean coast has begun building a new green neighborhood called Sociopolis, based on market gardens and ancient Moorish irrigation systems.

Other innovators are less exotic, but perhaps more important. In San Diego, garbage trucks run on methane extracted from the landfills they deliver to. Reykjavik is pioneering hydrogen-powered public transport. In Germany, they are greening the roofs of their high-rises to grow food, encourage birdlife, collect rainfall, and cool the streets below. Toronto air-conditions its buildings in summer with cold water from the depths of Lake Ontario. My city, London, charges most vehicles to come into the central area, a scheme intended to end gridlock and reduce its carbon footprint.

And the developing world isn't far behind. The southern Brazilian city of Curitiba pioneered bus-only roads, and then recruited the city's poor to recycle its garbage by offering groceries and bus passes in exchange. The air in two of south Asia's biggest and most polluted megacities, Delhi and Dhaka, has been transformed by switching tens of thousands of buses and motorized rickshaws to liquefied natural gas. You can taste the difference.

All this is important, but it's tailpipe tinkering. A more fundamental rethinking is needed. For the past century planners have designed cities as if resources like land, fuel, water, and concrete were unlimited, and as if waste was something to be dumped as cheaply and as distantly as possible. Cities need a new metabolism, conserving their resources, cutting our carbon-based energy sources, and mining their trash.

Let's start with the waste. The simple recycling bin should be a totem for the future city. Then, sewage needs rebranding as a feedstock, returning plant nutrients to the land to grow more food. And liquid effluent, suitably cleaned up, needs a makeover as the new drinking water. We must close the loop, recycling metals, paper, glass, plastic, and food leftovers.

It is common to think of rich countries as the recycling pioneers. But actually developing countries are generally better at it, because these resources are expensive, while labor is cheap. Chinese street sweepers get paid a tenth of the salaries they could expect in Europe; yet every aluminum can they pick up has a cash recycling value the same as in Britain. Whole communities live on landfills from Manila to Mexico City, sorting through garbage for materials fit for resale. We may want to improve their circumstances, but we cannot afford to lose their services.

But the key, I think, is to maximize the virtues of cities through good urban design. Well-planned cities can boost recycling, divert waste heat from power stations to heat buildings, tap sewage to generate bio-gas, and replace cars with buses, trams, and trains. The worst twentieth-century crime of urban planners was to design cities around cars. In the United States, the architect Frank Lloyd Wright provided a blueprint for modern America in his Broadacre City, a suburban idyll of homesteads connected by an endless lattice of highways. It became a global template, stretching from Milton Keynes in Britain (once heralded as an urban "autopia") to Brasilia, the modernist new capital that Brazil built in the bush in the 1950s.

The Frank Lloyd Wright generation "worshipped at the altar of the automotive god, and idealized mobility and freedom," says Peter Hall of University College London. He is a doyen among city planners, who did his own share of worshipping. Back then, the planners thought that people would in future have no desire to live in local neighborhoods. They would revel in the freedom provided by the car to pick and choose their neighbors from across the city. Christopher Alexander, professor of architecture at the University of California, Berkeley, saw neighborhoods not just as irrelevant, but as "military encampments designed to create discipline and rigidity." The subtext was clear: socialist neighborhoods bad, capitalist freedom good.

But the result of this infatuation was not total freedom, but gridlock, suburban blight, the ghettoization of people without cars, and burgeoning crime and social dislocation in bleak nightmarish urban landscapes. More *Clockwork Orange* than Broadacre City. Milton Keynes is a British national joke, and Brasilia is surrounded by the favelas it was supposed to replace. As a nondriver, quite at home in London, I despair at the sheer impossibility of getting around many other modern cities without a car.

So, back to the drawing board. We must unclog the cities' arteries by fighting the car's hegemony. Not just reducing its pollution and beefing up public transport systems, but transforming cityscapes to make the car as irrelevant as Alexander imagined the neighborhood to be. "Even if powered by biodiesel, hydrogen, or sunbeams, the private automobiles still require massive networks of streets, freeways, and parking structures to serve congested cities and far-flung suburbs," says Richard Register, founder of the campaigning organization EcoCity Builders in Oakland, California.

Eco-cities need to be a lot denser than Broadacre City. Sprawling cities like Houston, Texas, use cars ten times more than compact cities like Hong Kong, so some want a world of superdense Hong Kongs. Environmental futurologist Jesse Ausubel of Rockefeller University in New York envisages a twenty-first century of high-rise cities linked by underground magnetic levitation trains, and surrounded by expanses of restored countryside. But this eco-modernism sounds like a green nightmare—"another ruinous ideological fashion," as David Nicholson-Lord of the New Economics Foundation puts it. People don't want to live in machines. They want neighborhoods and access to green spaces and nature.

So is there a trade-off between eco-efficiency and pleasant living? Luckily the conflict may prove more theoretical than real. Supercompact cities may not be very eco-efficient after all. The closely packed cities in many parts of the world use less gasoline but run up huge air-conditioning bills to offset the heating effect from the dense forests of buildings. Stone, concrete, and asphalt absorb more solar energy, and reflect less, than natural surfaces such as grass, water, and trees. Air-conditioning systems themselves also give off heat, while tall buildings cut down the cooling winds. Superdense cities cook. In summer, Hong Kong uses more energy to air-condition buildings than for anything else.

This urban "heat-island effect" can be reduced by cutting direct sunlight through windows, increasing ventilation, cooling the air with water fountains and trees, even by painting buildings white. But the best solution is to get the urban density right: high enough for efficient public transportation, but low enough to minimize the heat island. Compact, low-rise cities such as London or Paris or Copenhagen seem to work best.

I hadn't realized that I have been living in an eco-city all along. And of course I have not. But London is a city better able to green itself than most. And now a group of architects dubbed the "new urbanists" want to create green cities that people will want to live in, by building neighborhoods with eco-efficient houses and low-rise apartment complexes, linked by wide boulevards with tramways down the middle and bicycle lanes down the side, with roadside cafés and parks and Metro stations all within walking distance. A bit like the nicer parts of London, in fact. Or Paris, almost everybody's favorite city. The eco-efficient city can also be a green and social place to live. Perhaps, says Hall, that is why rich people go green too. "The shift backwards to neighborhoods where the car is banished is happening most in the most affluent cities. You might say they can afford it. But it also shows that is what people want." He says once people have sufficient wealth and time, they go back to riding bikes and taking the train and sitting in cafés in pedestrianized city centers—and not having to drive everywhere.

But being green need not be a preserve of the rich. Another model of green living is, surprisingly enough, the shantytown. Shantytowns spring up by themselves, built by millions of people in the developing world, without a planner in sight. But they meet many of the ideals of eco-designers. They are high-density but low-rise; their lanes and alleys are largely pedestrianized; and many of their inhabitants recycle waste materials from the wider city. This is small-footprint living. And they have nooks and crannies that harbor wildlife. There may be sewage in the lanes, but there are few more diverse ecosystems than a ramshackle, chaotic favela. So perhaps something can be taken from the chaos and decentralized spontaneity of shanties and combined with a designer eco-city. That would be my ideal city.

The rethinking should go further. Twentieth-century planners saw no place in cities for growing food, or for untamed wildlife. Sure, they had "green lungs," but these were sanitized parks and golf courses. Some blueprints for eco-cities today have similar failings. But in past centuries, cities kept large areas of land close at hand to supply fresh vegetables. In London it was the Lea Valley. And less formally, urban agriculture is still alive and well around the world, as we saw in the last chapter. If we are all urbanites now, maybe we are all destined to be part-time farmers, too. And why not? Sometimes it will be done out of pleasure, sometimes out of necessity, and sometimes for profit. Simple market forces ensure that farms in cities get good prices.

Cities can also be havens for wildlife. I live close to the River Thames in London. All along the river, old industrial sites and their riverside wharfs have been abandoned. The derelict land is now being cleaned up and planted with upmarket apartment blocks. Greens love this. Friends of the Earth calls on the government "to act to allow polluted land to be cleaned up for the benefit of future generations." It demands that ministers set a target for 75 percent of new housing to be on "brownfield" sites. And the government is keen to deliver. One of the first promises from Gordon Brown after he became prime minister in 2007 was to build "eco-cities" on brownfield sites across the country.

But hold on. Less than a mile from where I live, a small brownfield site is being redeveloped. It used to be called Gargoyle Wharf, an oil-storage depot close to Wandsworth Bridge. Greens regarded the site as a toxic wasteland. But the plants thought differently. And for a while after the depot was demolished, it was London's top site for flowers. "We found more than three hundred species of flowering plants there," says Nick Bertrand of the London Wildlife Trust. "It was one of the most fantastic sites I've ever been to." But no environmental campaigners came to its defense. In the end, everyone decided to ignore the evidence of their own eyes, and development has gone ahead. Such is the power of the drive to redevelop brownfield sites.

Sheer madness, and born of the fact that we persist in believing that cities are environmental badlands, while the countryside is the exclusive home of wildlife. The truth is often the opposite. Many green fields that environmentalists want to protect are ecological deserts, says Peter Shirley of the Wildlife Trusts, whose members manage many of Britain's nature reserves. "It is not true that building on greenfield sites is always bad and building on brownfield sites is always good." But that is exactly what has been happening. This environmental philistinism has to stop. Cities need wild places as well as green places. The sanitizing of our environment can be an enemy to nature.

Cities are evolving in many ways. Often, say urban planners, they are ceasing to be conventional cities at all. They are merging into a type of wider urban landscape that the French geographer Jean Gottmann has dubbed "megalopolis." A big question is whether this development is good or bad for the greening of cities.

Megalopolis takes over when big cities become so unwieldy that they stop growing and instead spawn archipelagoes of urban centers around them. London has spread to create an urban region across southeast England that stretches west toward Reading and Oxford, north toward Cambridge, and now east along the Thames estuary. São Paulo is embracing a "golden urban triangle" that includes Rio de Janeiro and Belo Horizonte. The people of Mexico City have fled their congested and polluted megacity to surrounding cities like Toluca and Cuernavaca. Kolkata has dispersed across west Bengal. Tokyo is extending out to Japan's second megacity, Osaka, creating a megalopolis of 70 million people, linked by bullet train. Shanghai is joining hands with Suzhou, Nanjing, and Hangzhou, which will soon be just twenty-seven minutes away on a new maglev train with a top speed of more than 250 miles an hour.

Will megalopolis encourage or destroy the dream of eco-cities? It is easy to be gloomy. But, if properly designed, megalopolis could drive the creation of greener urban environments. The emergence of large urban zones with no single center could reduce commuting, for instance, because employment will be more dispersed. Likewise, losing the big central hub could give people a greater sense of living in neighborhoods. Tokyo has many intimate neighborhoods. In urban Europe people talk without affectation about living in "villages" within their cities. Some see urbanization as inevitably devouring the countryside. But maybe we need to rethink that distinction and bring wild things into the urban zone, which could embrace urban agriculture and green corridors more easily than conventional megacities.

Some say we can look back profitably for more ideas about how to create the cities of the future—to the walled medieval cities of Europe, such as Verona. Or Venice, the ultimate carless city that eco-planners can only dream of recreating. Or even the ancient cities of China. Suzhou, I noticed, is taking great pains to preserves its old walled gardens and waterways, even as it builds the brave new world beyond. I pondered this as I spent an afternoon with Ma Cheng Liang, a man in charge of redeveloping Shanghai in a greener mold, including building a new eco-city on an island in the Yangze delta. He outlined for me his many plans for "ecological modernism" in the city. It sounded a little bleak and technocratic. More about eco-efficiency than building a place people might want to live.

From Ma's office window, we could see below the huddled buildings and lanes of Shanghai's old town. As we talked, bulldozers were tearing down the buildings to make way for new office blocks. Ma saw this as progress. He wanted to do away with the past. Tear down the old town and create a bright, new, and green future. But to me, the old town was a dense, largely car-free enclave, mixing homes and workplaces and shops. It was a model of green design, the perfect embodiment of the dreams of the new urbanists. But he didn't see it. And still the bulldozers came.

Zero Carbon   28
Why We Can Halt Climate Change

The threat from global warming is greater than usually claimed. In my last book, *With Speed and Violence*, I explained why many scientists believe that the world faces a series of dangerous "tipping points" that could make warming happen much faster and more violently than allowed for in reports from the Intergovernmental Panel on Climate Change, the IPCC. The threats include the runaway release of greenhouse gases from natural reservoirs like soils, forests, and permafrost; the shutdown of the ocean currents and switch-off of the monsoon; and the rapid breakup of ice sheets, causing a sea-level rise of several meters within a century. All these could be unleashed as carbon dioxide levels in the atmosphere rise.

The IPCC is soft-pedaling on these doomsday scenarios in its most recent reports, because many of them are difficult to quantify and far from certain. But as one scientist put it to me, "If you were about to board a plane and the pilot told you there was a one in ten chance of it crashing, would you still take your seat and buckle up for take-off? Yet we are doing that with the planet." I don't know how close we might be to these tipping points. There are several decades of warming now "in the pipeline" that are unavoidable, so we may already have passed one or more. In that sense, it could be too late. But I'll stick with the leading U.S. climate scientist Jim Hansen, who thinks we may have a decade to turn the supertanker around. We could have longer, but you wouldn't want to bet on it.

Can we do it? Technically, absolutely. Can we do it without destroying our lifestyles? By and large, yes. As I hope the preceding chapter shows, we can have better, more fulfilling lives by adopting many green measures both personally and within the urban metabolism. And, as I argue in the next chapter, this may be part of our coming of age as postindustrial people. There is a plethora of technologies that allow us to use energy much more efficiently and stop generating that energy from carbon-based fuels: wind and nuclear, solar and biofuels, tidal and geothermal.

I am often asked "What is the 'greenest' form of fuel?" But I favor diversity. Biofuels are good—until we consume so much that we can no longer fill the boilers with rapeseed from local fields and start plundering palm oil grown in the rain forests of Borneo. Wind power is good unless we cover every hilltop in the country with turbines. Tidal and wave power are good, but there would be an outcry about the pylons needed to bring serious megawatts from the best sites on the coasts, through picturesque landscapes to distant urban markets. All these energy sources are good, but in moderation.

What about nuclear? There is a conundrum here. If someone offered us the chance to use a tried and tested technology to produce up to half our electricity with very low carbon emissions, only a small impact on the environment, and with no requirement to rewire the grid, surely we would jump at the chance. But they did, and so far we haven't. It is nuclear power.

I have never liked nuclear power—less because of safety worries and more because I fear nuclear proliferation and the erosion of civil liberties needed to keep the technology safe. But in the face of the threat from climate change, I find it difficult to maintain my opposition. In general, I think the safety fears have been overblown by objectors to nuclear power. And it seems to me that, in the wake of 9/11, we have already lost many of those cherished civil liberties. I am quite certain that no more nuclear power stations should be built until governments have finally (after half a century of failure) come up with solutions to the big technical problems about what to do with nuclear waste. But I fear—and I do mean fear—that we may need this in our armory as well.

Coal? It produces more carbon dioxide for a given amount of energy than any other major fuel. But maybe we need even that, if we can develop the technology to bury its emissions out of harm's way. Not otherwise, however. Coal should be banned otherwise. Everybody talks about hydrogen as the fuel that could solve all our problems. But it is not a source of energy in itself. It is just a way of storing and transporting energy. Before we can burn it we have to make it. And making hydrogen is a very energy-intensive activity. So tell me how you want to make your hydrogen and I'll tell you if I am in favor.

Maybe the future is not in centralized forms of energy generation at all. Some think that smart electricity grids of the future will allow us all to generate electricity in our backyards by any means, and sell to the grid. As a way of storing the intermittent energy from wind power and the like it could be a boon. Maybe a new generation of cheap solar cells will allow us to make our buildings out of them. Then our roofs and walls and windows could soak up the sun to give us power.

Most of these technologies are economical, with oil now peaking at $100-plus a barrel. Some are not yet quite as good as they seem, because they are themselves quite energy intensive—think of the energy needed to make fertilizer to grow the biofuels, to build PV cells or refine the uranium fuel for nuclear power. That is one reason I would give priority to energy efficiency measures. They are cheap, plentiful, and often save money in the very short term. In many areas of life we can make huge improvements by simple and painless changes to the way we live our lives. The one area where there are no technical fixes at hand is air travel, which is the biggest source of emissions from many people with the biggest carbon footprints, including me. We simply have to give up flying as much as possible.

I don't have a magic formula for saving the world. The neatest idea I have heard is to combine technology for capturing and burying CO_2 with burning biofuels. Biofuels soak up carbon dioxide from the air as they grow. So if you run a power station on biofuels and then bury the carbon emissions, you have built a system for sucking CO_2 out of the air and burying it back underground. That way we could even start lowering CO_2 levels in the air. And I think it makes a lot more sense than nutty schemes like shading the planet beneath a huge parasol put into space. Of course the biggest question with biofuels is whether we have the land and water to grow them as well as food. But it is an option.

Mao used to say: "Let a hundred flowers bloom; let a hundred schools of thought contend." Maybe that is right. There are almost too many options. Pick your own. One thing I am certain about. I just hate knowing that more of my electricity comes from Drax than anywhere else.

We may have to change things fast to avoid tipping points in the climate system. But maybe we will. For there could be tipping points in our political and economic and cultural systems, too. I do detect big changes afoot. Many large corporations, especially in Europe, are now actively asking for governments to set tougher targets on greenhouse gas emissions. They believe they are the quickest, smartest kids on the block, and that they can gain a market advantage over their slower rivals once the limits are in place. Not all corporations are like that, of course—and even the best will only act under strong public pressure. Yet I believe there is the potential for a "race to the top" based in part on the profit motive. But if it is to happen, it will be because we, as customers and voters, demand it.

Some say that there is little point in the rich world getting its act together if developing countries like China, India, and Brazil don't act too. They point out that China builds two coal-fired power stations a week and in 2007 became the world's biggest CO_2 emitter. Ultimately, that argument is correct. But remember this. China is the world's most populous country; you would expect it to have high emissions. Looked at per head of population, its emissions are much lower than ours. And it has been emitting them for much less time, so the amount of carbon dioxide in the atmosphere that came from China is much less than, say, from the United States and will remain so for many years. If Westerners like me had carbon emissions like those of the average Chinese, there would be no climate change problem right now.

Remember too that industrializing countries will want to manufacture energy-efficient goods to sell to the rich world. They will want to invest in the new energy technologies, not the old ones. And, increasingly, they see climate change as a threat to their own economic development. The Stern Report, published in 2006, on the economic implications of climate change made that abundantly clear.

And by changing the language to recast climate as an economic rather than a scientific or environmental issue, Stern has galvanized the world's political classes in a way that scientists and environmentalists failed to do.

My bet, at least on my more optimistic days, is that the whole world will soon be embracing a move to a low-carbon economy. It may happen with or without internationally agreed emissions targets, though I think that tough targets would make it happen faster, by increasing the economic incentives to cut emissions. The next generation may look back and wonder what the fuss was about. Maybe. I cannot guarantee it will happen. And it certainly won't happen if we sit back and assume that it will. Nor can I guarantee that it will happen before some speedy and violent tipping point in the natural world engulfs us.

Actually my greatest concerns are less about the application of the technologies to fix climate change, which are there if we will only use them, and more about the winners and losers along the way. That is why I think the argument about Kenyan bean farmers is so important. If climate change becomes an excuse for environmental protectionism, then, frankly, I hope we all fry. We have to cut our carbon emissions in ways that do not impoverish the poorest—those whose personal carbon emissions are the lowest. We need fair-traders, not green patriots. We need to maximize our positive social footprints as well as to minimize our negative ecological ones. But we can do it.

Defusing the Bomb   29
Why We Can Halt Population
Growth—and Save the World

Some people see global population growth as the ultimate driver of our environmental predicament. There are, they say, just too many people with too many footprints. I don't want to be a people hater, and I don't see people as "pollution." We are not born with original environmental sin. I am a humanist, and I think our problems today come not so much from too many footprints as from the much smaller number of large footprints. Like mine. Overconsumption rather than overpopulation is killing our planet. As Gandhi put it: there is enough for everyone's need, but not for everyone's greed. I also believe the human footprint can be good as well as bad, and we have to strive for that. But I grant you the strain of almost 7 billion footprints is profound.

Now for the good news. We can halt population growth. And we have a savior. Her name is Isabella, or maybe Clara or Bianca—emancipated Italian women on childbirth strike. Thanks to them and a growing sisterhood round the world, the baby boom that the world has feared for half a century could be turning to a baby bust. The planet's population could be in long-term decline within another fifty years. Future generations may worry that their family tree is dying. But let's start with Isabella and her friends. Isabella lives in Rome. She is in her thirties. She has a nice apartment, a nice job, and a nice boyfriend who still lives at home with his mother. So Isabella can see him when she likes. The last thing Isabella needs is children. She would probably have to give up the apartment. Her employers would swiftly find a younger model—and so might her boyfriend. Even if they did stay together, he would have to move in with her. He would leave the place a mess, spend her money, and be hopeless at changing diapers.

It is for such reasons that the women of Italy are making very few babies. With splendid irony, the home to the Catholic Church, with its fundamentalist opposition to artificial birth control, has the lowest fertility rate in the world. At 1.2 babies per woman, it is far below what is needed to maintain the current population.

And Italian women are not alone. In Spain, Greece, the Czech Republic, and a veritable collective of former Soviet states such as Russia, Lithuania, and Armenia, their sisters have fertility rates of 1.3. These are probably the lowest rates in nonsuicidal populations in history.

All this is quite a turnaround. For half a century, the world has been scared by forecasts of exponential population growth. One of the cornerstones of environmentalism has been that this demographic deluge will overwhelm the supply of resources such as food and water, while triggering rampant pollution and global warming. Such fears are not surprising. During the twentieth century, the world's population increased almost fourfold, from 1.6 billion to 6 billion. We passed 2 billion in 1927; 3 billion in 1961; 4 billion in 1974; 5 billion in 1987; and reached 6 billion in 1999. Every additional billion came quicker than the last. The reason was simple: falling death rates, especially among children, combined with continued high fertility.

The deluge is not over. Every year the world has 80 million more inhabitants. But we have now passed the point of maximum annual increase in numbers. It took twelve years to get from 5 to 6 billion; but it will take fourteen years to get to 7 billion, and perhaps three decades to reach 8 billion. This is no statistical blip. Almost since we started fearing the baby boom, women have been having fewer children. In 1950, worldwide the average woman had five children. Today she has just 2.6.

The rule of thumb among demographers is that a stable long-term population in the modern world requires each woman to have 2.1 children. The extra 0.1 is expected to compensate for girls who do not live long enough to have families. So global fertility is clearly fast falling to replacement levels. The sheer number of women of childbearing age means that to crank down population growth we have a long way to go. But as the twentieth century baby boomers age, the tide will turn. The population time bomb, the subject of a million millennial nightmares, is steadily being defused.

Looked at historically, our recent demographic surge is not so unexpected. Over time, population growth has been concentrated in a series of surges, generally following technological and cultural revolutions. The first revolution came with the more sophisticated development of tools, seemingly following a giant volcanic eruption of Mount Toba in Sumatra, which almost wiped out *Homo sapiens* some seventy thousand years ago. The second revolution was the spread of agriculture and stable settlements from the Middle East after the end of the last ice age, ten thousand years ago. The third revolution was the industrial revolution, which is still working its way around the world today. It has raised the world's population to 6.5 billion already. The key features of this revolution have been advances in food production, which have allowed the world to feed more people, and new methods of controlling diseases, which have allowed people to live longer.

The sheer scale of what has happened in the past century or so is without precedent. There seems little doubt that the twentieth century will stand out, perhaps for millennia to come, as a century of unique explosive growth in the human population. *Homo sapiens* —naked apes with attitude and big ambitions—were at their most fecund. But demographers have long assumed that this great population surge will fizzle out during the twenty-first century, as most of the world's women settle down to a conventional Western family life with mother, father, and two children.

They call this boom followed by stabilization the "demographic transition." It described well enough the progress until recently of the first countries to industrialize. In Europe the annual death rate declined from more than thirty deaths per thousand people in the seventeenth century to below ten today. With fertility rates remaining high, population growth rates in Europe at the start of the twentieth century reached a peak of around 1.5 percent a year. Then they began to fall, as people responded to the greater survival rate among children by having fewer babies. In many developing countries, the transition has been telescoped into a few hectic decades, during which national population growth rates have often exceeded 3 percent per year, and populations have doubled in less than twenty-five years. But the pattern remains the same.

All according to plan. Except that nobody told women about the demographers' endgame, in which fertility rates would settle down again at replacement levels. And there is a growing suspicion that Isabella and her sisters are leading the way to a different demographic future. Based on current trends, the world's population is primed to start diminishing for probably the first time since the Black Death in the fourteenth century. Already, more than sixty countries have fertility rates below replacement levels. The club now includes much of the Caribbean, Japan, South Korea, China, Thailand, Sri Lanka, Iran, Turkey, Vietnam, Brazil, Algeria, Kazakhstan, and Tunisia. Other large developing countries, whose populations have soared in the past half century, expect their own national fertility rates to fall below replacement levels within twenty years. They include India, Indonesia, and Mexico.

A few of these countries have brought birth rates down through coercion. China's one-child policy is the most notorious. But that is not the situation generally. Often birth rates have plunged despite opposition from governments and religious leaders. Catholic influence has meant that the Brazilian government has been slow to develop state family planning. Even so, a third of married Brazilian women have been sterilized, and fertility has more than halved in twenty-five years, to 1.9 in 2005.

The case of Iran is even more remarkable. Despite a period of fundamentalist rule by the mullahs, fertility rates crashed from 5.5 in 1988 to just 1.8 in 2005. In late 2006, President Ahmadinejad reiterated the old mullahs' call for a homegrown baby boom, and for women to return to their "main mission" of having babies. But he seemed out of line with both the clerics, who now back birth control, and his own administration, which runs a condom factory.

Some argue that countries stuck in poverty will buck the trend.

But in practice, some of the world's poorest countries are emptying the maternity wards. Bangladeshi girls are among the least educated and most likely to marry while in their early teens. Yet they give birth to an average of just 3.1 children, half the number their mothers did. The garment workers in Dhaka and the poor women in the villages around Khulna told me without exception they were determined not to have more than two or three babies. Around half of them use contraception routinely. Provision is free, but there is no hint of compulsion.

Rich or poor, socialist or capitalist, Muslim or Catholic, secular or devout, with tough family-planning policies or none, most countries tell the same story: couples are engaged in a coital revolution. In Japan the condom is tops, in China the IUD is king, in Latin America it's sterilizing and in Europe the pill. Turkish men are reputed masters of coitus interruptus, while more trigger-happy British males opt for vasectomies in record numbers. But however it is done, babies are out.

Why is this? Some point to urbanization. In poor rural societies, children are vital as labor in fields. But in cities, children cease to be economic assets and instead become economic liabilities—costly to educate, clothe, and feed. People invest in material goods rather than children. Jack Caldwell of the Australian National University in Canberra, one of the doyens of demography, goes a step further. He argues that female emancipation is the natural social consequence of the shift from agricultural societies, built around the family unit, to the modern industrial world. "Postagricultural society does not need the traditional family," he says. "At the level of the individual, there is no necessity for either families or fertility. The individual has been freed." In other words, it is the logical, but previously unseen, conclusion of the demographic transition.

Perhaps so, but needs and wants can be different. Women may have children because they want them, not because they need them. The real answer may be simpler. Young women may no longer want to be wives and mothers. Tim Dyson, professor of population studies at the London School of Economics, says not having children has become a statement of modernity and emancipation. The spread of

TV in particular has opened women's eyes to a whole new world, and modern birth-control methods are helping turn some of their aspirations into reality. In the modern world it is not so much literacy that leads to lower birth rates (though that certainly helps because it is a practical aid to getting on in the world), but more the flickering screen in the corner of the room that shows young women the world they could have if they could win a life away from rearing children.

Motherhood is an increasingly alien activity for many working women. As *Guardian* columnist Madeleine Bunting put it, for those who take the plunge "motherhood hits most women like a car crash. Nothing in our culture recognises, let alone encourages, the characteristics you will need once a bawling infant has been tenderly placed in your arms." And women have too much else to do. "Go to rural India," says Dyson, "and you find that women are fed up with the men, who seem to be going nowhere. It is the women who are increasingly running the farms. It is the women who are getting jobs and taking charge. They don't have time to have children anymore." He exaggerates. But there is a new assertiveness in young women across the world that transcends the rural–urban divide.

We should not forget those parts of the world where fertility rates are not in freefall. The list is revealing. Of the thirty countries with the highest fertility rates, where five- and six-child families are still the norm, twenty-six are in Africa, where death rates are often rising because of AIDS. Three of the others are Muslim states—Afghanistan, Yemen, and Oman—where most women remain relatively ill educated, in the home, and cut off from outside influences. But these seem to be holdouts. Fertility is falling fast now in Pakistan and Saudi Arabia. Where is all this heading? Will the rest of the world's women follow the lead of Isabella and her friends in Italy? There is an alternative model. In Sweden, I met Astrid. She has two children and got a year's maternity leave when each of them was born. She works a flexible thirty-hour week and can use daycare at the office when she needs to. Her husband is an expert diaper changer and shares the 4 a.m. feeding duties.

Not surprisingly, Astrid and her friends feel more able to have a family than Isabella's crowd, who have little chance of getting part-time jobs or securing places in daycare for their children. In consequence, Sweden's fertility rate is 1.7 children per woman. That's not enough to maintain the country's population in the long term, but neither is it an Italian-style demographic meltdown. Most northern European countries have maintained higher fertility rates than countries around the Mediterranean. Norway's is 1.8; Britain and Finland notch up 1.7. The most fecund women in Europe today are the French at 1.9. Meanwhile, in Eastern Europe fertility rates have plunged since the collapse of communism wrecked state-funded support services for families. In Soviet times, Olga worked in a university laboratory in Moscow, where child care was rudimentary but universal. Her family badly needed her money to make ends meet, but at least she could pursue a career. Now she says that the new private employers are not interested in providing child care and her daughter will not be able to have children and a career.

So will the rest of the world follow northern Europe, or its neighbors to the south and east? Isabella or Astrid? Brigitte or Olga? Caldwell thinks the signs are clear. "Italy is the future. The Mediterranean patriarchal model is far more common in the world than the northern European model of more helpful husbands." Conservative family values lie behind the ultra-low fertility rates among the office girls of Shanghai and Tokyo. If he is right, then a demographic implosion is starting to unfold. If current trends persist, we will begin to see absolute falls in the indigenous populations of most industrialized countries. Italy and Hungary expect a 25 percent drop in population, and Japan 30 percent. Eastern Europe could lose between a third and a half of its populations by midcentury—a trend exacerbated by the "vodka effect," which has cut male life expectancy in Russia to fifty-eight years. By 2050, Russia may have fewer people than Uganda. By 2100, Britain's indigenous population can be expected to halve, while Italy's population could crash from 58 million now to just 8 million. Germany could have fewer people than today's Berlin.

Without a sharp rise in fertility rates, only mass migration into Europe can halt this. Perhaps that is how it will play out. The new demography may create a planet of itinerants as labor becomes globalized along with capital. We already see Indian workers building

Dubai, Poles in the fields of England, Chinese foresters in Siberia, Guatemalan grape pickers and housemaids in California.

But where will the new migrants come from if global populations start to fall? Demographers are reworking their predictions for future world populations. If the world moves toward an average of 1.85 children per woman, world population would peak at about 7.5 billion around 2050, and then fall. By 2150 there would be 5.3 billion people on the planet. But 1.85 children looks increasingly like a rather high estimate. Most countries are heading for much lower figures. If women settled for even a Swedish-style fertility level, we would be down to 3.2 billion by 2150—roughly half today's population. What's the big deal? A decline in the world's population sounds like a good thing. It would reduce the pressure on the planet's natural resources. The air would be cleaner, biodiversity richer, soils more fertile, climate more predictable, and pollution less toxic. But a world with a falling population will be different in other ways that might be less agreeable. For one thing, the world's population will be very much older. The average age of a world citizen is around twenty-nine years today. By 2050, it will be over forty, with a fifth of the world's people over sixty. The average Japanese or Chinese will be over fifty.

A grayer world might be less innovative and more conservative. More boring, perhaps. It would be a world in which the richest countries compete for diminishing supplies of migrants. Eastern Europe, which is already suffering what the World Bank calls a "transition from red to gray," may simply curl up and die.

So is demographic decline a bad thing? When I first wrote about this, I got a splendid letter from California historian Theodore Roszak. "I can't recall an exercise in social analysis more wrongheaded," he wrote of my article. "On what grounds do you assume that something like our current human numbers are optimum?" I don't, as it happens, believe in any optimum number. But his main point was to sing the praises of the old. "When there are more people above the age of fifty than below," he wrote, "the longevity revolution may at last offer industrial civilization the preconditions for sustainability." We will be older, but wiser. That point I certainly take. And I concede it took an older man to point it out. We are coming through the greatest surge in human population in our history. And it has the potential to alter the psyche of our species. By the end of this century, *Homo sapiens* could well be more conservative, less innovative, more boring than we were in the twentieth century. But that may be no bad thing. A stable, sagacious society that has thrown off its current adolescent restlessness and settled into middle age sounds appealing. The tribal elders may take center stage once more. We might look after our planet much as fastidious middle-aged couples look after their homes.

But we do need children, too. And there is something wrong with a society that doesn't want to produce them. There may be nothing wrong with a century or so during which fertility rates fall to 1.6. There would be a great deal wrong with a global fertility rate of 1.4 or below.

As countries seek to revive their childbearing resources, they may flirt with draconian measures, cutting women out of the workforce and keeping them at home, banning abortions, and restricting access to family-planning services. But that is unlikely to work. Women won't stand for it. Instead, Dyson says we need a continuation of the "renegotiation" of gender roles under way across most of the world, and most advanced in northern Europe. Paradoxical as it may seem, the female emancipation that has reduced family size so dramatically in the past fifty years will need to be extended rather than dismantled if family sizes are to rise from the worst-case Italian model. It is the incompleteness of emancipation that is the problem.

As women grab their new freedoms, the most pressing need is to instill new responsibilities in men—and in the state. In most of the world today, fertility is plunging far below replacement levels because women have decided they want to become more like men. Right now that leaves little room for babies. To change that, men must take the plunge and start to become more like women: better housekeepers, better child-rearers—and perhaps better custodians of the planet, too. The future of *Homo sapiens* could depend on it.

The twentieth century was tumultuous. It may go down in history as a golden era of human inventiveness, but also of human folly. The party to end them all. But the twenty-first century is the century when, in the cold light of dawn, we discover if our species has the resources to live with the consequences, and put right the damage. We face the most almighty hangover, as the toxins unleashed by our century-long binge work their way through the earth's system. We have to detoxify. We have to sober up. We have to come to grips with exactly how much damage we have done to the planet, its land and oceans and atmosphere, and figure out what it will cost to repair it. We have, in short, to grow up. Luckily, as a species, we will be growing older. A time for responsibility, we can hope. So this is not elaborate metaphor, but more or less literal truth.

As we grow up, we have to feminize our society, green our society, and become better custodians of the planet. There is no guarantee we will make it. We have no reason to believe that our luck will hold. We may already have done too much damage and be doomed to see our civilizations wrecked by climate tipping points, and our numbers crashing through disease and hunger and war. But I do believe we still have the ability, and maybe the foresight, to mend our ways. Politicians talk, apparently seriously, about reducing greenhouse gas emissions by 80 percent and more by midcentury. Global corporations are responding to real market pressures when they declare that they are going carbon-neutral and selling organic and fairly traded and environmentally certified produce. They have corporate social responsibility departments because they need them. There may be some cynicism involved, some greenwash. But I do believe that they increasingly recognize that these are jobs that have to be done.

There will be false dawns and false starts. Maybe putting a tiny wind turbine on your roof is a silly idea. Maybe it makes more sense to incinerate your old newspaper and make some energy rather than drive it across the country for recycling. Maybe mass-produced biofuels are a dead end. Maybe we are kidding ourselves sometimes when we buy fair-trade products and genuinely believe that they are fairly traded. But we should not make the perfect the enemy of the good. And if our green and ethical initiatives don't always have the desired effect—as they certainly won't—then we have to build on the occasions when they do. Buy the organically grown, fairly traded cotton, socks, and coffee.

I think we need to remember the personal, as well as spending our time calculating carbon footprints. My journey for this book was about people as well as my environmental footprint. I cherish meeting the Milonge brothers recycling my old clothes, and Ranabhai, who knew exactly how much better off he was growing organic cotton, and the Wonder Welders, and Ado with his sheep in Kano, and many others. I hope that some of the girls on the production lines of Suzhou get to become "queens of trash" like Cheung Yan. I wonder whether Aisha and Miriam and Akhi will go back to their villages or stick it out in the city. And whether George Tagarook will get his new fire engines, and Sammy Cheng is still churning out mice for the world. These small pictures help make the big picture. Just as my footprint is part of the global footprint of humanity.

In every good story there is a moment of high drama. A moment when futures are sealed. When events turn on a twist of fortune or some trick of fate. When you are part of the story, it is probably hardest to know when that moment arrives. Probably most generations think that theirs is special. But we do seem to be close to the moment when the fate of *Homo sapiens* may be sealed. When we can choose to follow our good or our bad instincts.

The oldest known human footprints are in the mud of Laetoli in Tanzania. They were left more than 3 million years ago by a species of humans long since extinct. We, *Homo sapiens*, have been clever enough to find and understand those footprints. No previous species on the planet could have done that. It would be a shame if our cleverness were to be our undoing.

Sources and Acknowledgments

I have resisted the temptation to pepper this story with footnotes and references. Generally, like any journalist, I have indicated my sources in the text, whether they are personal observations, direct conversations, or printed and online sources.

This was a long journey and many people provided me with personal help along the way. Some don't even know they helped; others would rather you didn't know; and still others I have, no doubt, failed to remember as I write this. But thanks anyway to the following:

At Goldfields, Willie Jacobsz was keen to get me underground, and Rob Chaplin and Melville Haupt kept me safe and informed while I was down there. On Mount Kilimanjaro, John Weaver of Twin Trading was a most assiduous guide, while Simon Billing back in London swept up my later queries. Raymond Kimaro, general manager of KNCU, was welcoming, and, at Cafédirect, thanks go to Katya Bobova. Harriet Lamb at the Fairtrade Foundation offered great advice and assistance at the start of my research.

WWF took me to Mauritania, Alaska, and Chengdu in China. In Mauritania, park ranger Antonio Araujo introduced me to desert fishers, while Pierre Campredon and Luc Hoffmann talked good sense. In Alaska, thanks especially to Denise Meredith and Pam Miller.

Iqbal Ahmed and Alec Dodgeon at Seamark may not be happy with my take on their industry, but I thank them nonetheless. Tutu, Dadu, Quazi Mukto, and Abu Hasan Bakul all helped me in the heat of Khulna. In Dhaka, Khushi Kabir, Nazma Akter, and, especially, Khorshed Alam helped me with both prawns and the rag trade. Thanks also to Bob Pokrant, Bill More, Juliette Williams, and many others.

Stefano Padulosi of the International Plant Genetic Resources Institute, now renamed Bioversity International, has been a fount of information on wild and orphan crops of all sorts. Emile Frison provided invaluable support for my research into bananas, as did Rodomiro Ortiz on an earlier trip to his research station in the Niger delta, run by the International Institute for Tropical Agriculture. That trip, organized by Anne Moorhead of the IITA, also took me to Cameroon to explore cocoa farming in that country, where Stephan Weise was also a great help, and to Kano, where I heard B. B. Singh's views on feeding Africa.

In Kenya, Michael Mortimore, an independent-minded Africa hand in the best tradition, introduced me to Machakos. Dicky Evans and Richard Fox facilitated my visit to Homegrown in Kenya, and John Simeoni and Rod Evans took me around.

In Australia, Tom Burke and Rick Humphries of Rio Tinto gave me the run of Gladstone. And, during the same visit down under, Kelly Chapman and Dan Hickey took me out to the cotton farms during a break from the Brisbane River Symposium. Thanks also to Allan Williams.

At M&S, Graham Burden did his best to help me understand the cotton business, and Mike Barry was candid on corporate strategy for sustainable development. At Agrocel in India, Hasmukh Patel and Dilip Chhatrola were invaluable. And later, at Maral and Arvind, Amit Baid and Yash Baheti, respectively, could not have been more helpful. I visited Uzbekistan with a grant from the International Water Management Institute, whose Tashkent officer, Iskandar Abdulaev, accompanied me to Karakalpakstan.

Logitech's Ben Starkie fixed me up with Sammy Cheng and his friends in Suzhou. Xiufang Sun talked me through the Chinese timber trade, with help from Greenpeace and from Scott Poynton of the Tropical Forest Trust and his colleagues.

Michael Renner of the Worldwatch Institute helped with war and the "resources curse," as did Arjen Hoekstra on the virtual water trade and Tricia Feeney on the coltan scandal. Thanks to Sten Nillson at the International Institute for Applied Systems Analysis in Austria, along with his Siberian associates, for taking me to

Noyabr'sk. But no thanks to Gazprom and its agents for curtailing my visit to Novy Urengoi.

Midwinter Productions arranged for my journey on the Thames waste barge. Ray Georgeson and Doreen Fedrigo guided me through the recycling maze. Paul Llewellyn introduced me to the virtues of sewage on beaches. Lynne Wright introduced me to the Milonge brothers in Dar es Salaam. Thanks also to the people of Cookshop and the estimable Paul Joynson-Hicks for finding a new owner for Joe's phone in that same city.

I visited Chinese recyclers thanks to Robert Gibson and colleagues at Swire in Hong King and Xiamen. Priti Mahesh and Ravi Agarwal showed me the horrors of Mandoli, and Tony Roberts of Computer Aid International introduced me to Nairobi computer refurbishers.

Peter Hall and Jesse Ausubel told me much about urban planning, green and otherwise. Thanks, as always, to the two Jims, Lovelock and Hansen, for their erudition on climate change and the science of our planet. Jack Caldwell was, similarly, a guru on demographics.

Parts of this journey, and much of my background knowledge, would not have been possible without the support of Jeremy Webb and the commissioning editors of *New Scientist* over many years. Some readers of that magazine will recognize snatches of material recycled here, particularly from my continuing blog, *Fred's Footprint*, to be found on its website at www.newscientist.com.

My UK editor, Susanna Wadeson, and agent, Jessica Woollard, latched on to my idea for this book with enthusiasm, and saw it through. At Beacon Press I am indebted to many people, most notably Amy Caldwell, for their editorial intelligence and publishing integrity and enthusiasm.

And thanks finally to my family—the sometimes mysterious *we* in the text—who sustained the household on which this narrative is based and cajoled me in a green direction that I have often resisted, through laziness or a desire not to pollute this "warts and all" investigation. Now, of course, I have no excuse.

Index